T0299721

Communications System Laboratory

Communications
System Laboratory

B. PREETHAM KUMAR

CALIFORNIA STATE UNIVERSITY,
SACRAMENTO, USA

CRC Press
Taylor & Francis Group
Boca Raton London New York

CRC Press is an imprint of the
Taylor & Francis Group, an **informa** business

MATLAB® and Simulink® are trademarks of The MathWorks, Inc. and are used with permission. The MathWorks does not warrant the accuracy of the text or exercises in this book. This book's use or discussion of MATLAB® and Simulink® software or related products does not constitute endorsement or sponsorship by The MathWorks of a particular pedagogical approach or particular use of the MATLAB® and Simulink® software.

CRC Press
Taylor & Francis Group
6000 Broken Sound Parkway NW, Suite 300
Boca Raton, FL 33487-2742

© 2016 by Taylor & Francis Group, LLC
CRC Press is an imprint of Taylor & Francis Group, an Informa business

No claim to original U.S. Government works

International Standard Book Number-13: 978-1-4822-4544-8 (Hardback)

This book contains information obtained from authentic and highly regarded sources. Reasonable efforts have been made to publish reliable data and information, but the author and publisher cannot assume responsibility for the validity of all materials or the consequences of their use. The authors and publishers have attempted to trace the copyright holders of all material reproduced in this publication and apologize to copyright holders if permission to publish in this form has not been obtained. If any copyright material has not been acknowledged please write and let us know so we may rectify in any future reprint.

Except as permitted under U.S. Copyright Law, no part of this book may be reprinted, reproduced, transmitted, or utilized in any form by any electronic, mechanical, or other means, now known or hereafter invented, including photocopying, microfilming, and recording, or in any information storage or retrieval system, without written permission from the publishers.

For permission to photocopy or use material electronically from this work, please access www.copyright.com (http://www.copyright.com/) or contact the Copyright Clearance Center, Inc. (CCC), 222 Rosewood Drive, Danvers, MA 01923, 978-750-8400. CCC is a not-for-profit organization that provides licenses and registration for a variety of users. For organizations that have been granted a photocopy license by the CCC, a separate system of payment has been arranged.

Trademark Notice: Product or corporate names may be trademarks or registered trademarks, and are used only for identification and explanation without intent to infringe.

Visit the Taylor & Francis Web site at
http://www.taylorandfrancis.com

and the CRC Press Web site at
http://www.crcpress.com

To Veena, Vasanth, and Alex

and

In memory of my parents

Contents

Preface

The idea behind this book and its structure originated primarily from our students, who had often expressed their interest and desire to read background theory explained in simple terms and to obtain practical computer and hardware training in the relevant field. This motivation helped in the production of our previous book titled *Digital Signal Processing Laboratory*, and the current book follows that trend. The intended audience are primarily undergraduate students taking a course in communication systems for the first time. The book is highly relevant at the present time, with rapid advances in wireless and cable communications, and it is vital for the students to complement theory with practical software and hardware applications in their curriculum.

This book essentially evolved from the study material of two courses taught at the Department of Electrical and Electronic Engineering, California State University, Sacramento (CSUS). These courses, *Modern Communication Systems* and *Communication Systems Laboratory*, have been offered at CSUS for the past several years. Students who took these courses gave very useful feedback, such as their interest for an integrated approach to communication systems teaching that would comprise a parallel training in theory and practical software/hardware aspects of this sometimes challenging subject material.

Hence, based on their feedback, each chapter of this book has been designed to include the following components: a *brief theory* that explains the underlying mathematics and principles, a *problem solving* section with a set of typical problems to be worked by the student, a *computer laboratory* with programming examples and exercises in MATLAB® and Simulink®, and finally, in applicable chapters, a *hardware laboratory* with exercises using test and measurement equipment.

In Chapter 1, we go into a brief review of wired and wireless communication technologies and how the world is connected by using these different systems. This review is followed by a computer lab, which introduces the students to basic programming in MATLAB and creation of system models in Simulink. This chapter concludes with a hardware section, which contains instructions and exercises on the usage of basic signal sources, such as synthesized sweep generators, and measuring equipment, such as oscilloscopes and spectrum analyzers.

Chapter 2 focuses on time and frequency analysis of communications signals and systems. Specific topics include Fourier transform theory and practical computational methods like the FFT. This is followed by a computer lab, which involves time and frequency simulation of basic signals and systems. This would form the basis for the simulation of practical communication systems in later chapters. This chapter concludes with a hardware laboratory, which includes the measurement of signal spectrum using signal analyzers and spectrum analyzers.

Chapter 3 covers the *first-generation* analog modulation systems, which is an important introductory material for students living in a digital age. The theory section covers amplitude and angle modulation systems, and the software lab includes Simulink modeling of these systems. The hardware lab involves actual fabrication of simple AM and FM circuits and time/frequency measurement of these circuits.

Chapter 4 moves on to *second-generation* digital communications systems, covering the analog-to-digital (A/D) process, and digital modulation techniques such as Phase Shift Keying (PSK), Frequency Shift Keying (FSK), and Differential PSK systems. The software lab includes Simulink model simulation of these digital communication systems and their time/frequency characterization. The hardware lab gives guidance on the construction of simple shift keying systems and measurement of their performance.

Following a similar evolutionary process, Chapter 5 focuses on the *third-generation* (3G) Spread Spectrum technology, including Frequency Hopping and Direct Sequence spread spectrum systems, which forms the basis of several international standards such as CDMA2000, W-CDMA, TD-SCDMA, and Wi-Fi. The computer lab provides training to students in the simulation of the latter practical 3G systems, using Simulink modeling.

Chapter 6 extends the capacity of broadband communication systems to the *fourth generation*, with the discussion on systems such as 4G and 5G Long-Term Evolution (LTE). Since these are rapidly evolving technologies, the software lab focuses on addressing challenges with the high-speed gigabit simulation of MIMO (multiple input multiple output) systems and testing solutions such as frequency-scaled processes.

Finally, Chapter 7 includes a brief discussion of emerging trends in Wireless Local Area Networks (WLANs) and Ultra Wideband (UWB) systems, and the vision behind the development of future wireless networks.

There are six appendices. The first four appendices give detailed hardware descriptions and user instructions for the equipment used in this book. Appendices A, B, C, and D cover four equipment models, respectively: synthesized sweep generators, spectrum analyzers, dynamic signal analyzers, and digitizing oscilloscopes. Appendix E gives typical examples of practical integrated circuits used in communications technology. Finally, Appendix F illustrates the spectrum allocations around the world.

I sincerely thank the many people who have contributed directly and indirectly in bringing this effort to fruition. I sincerely appreciate the encouragement from the CRC Press publisher Nora Konopka, for her continued confidence in me to produce this second book. I also thank Jill Jurgensen, project coordinator at CRC Press, for steadily guiding me along during the actual preparation of the chapters. Most importantly, I would like to thank all the students at CSUS for their important feedback on the communication system courses, which formed the basis of this book.

Finally, I am indebted to all my family and friends for their support and inspiration, above all, my wife, Priya, who continues to encourage and motivate me in all my endeavors.

MATLAB® is a registered trademark of The MathWorks, Inc. For product information, please contact:

The MathWorks, Inc.
3 Apple Hill Drive
Natick, MA 01760-2098 USA
Tel: 508-647-7000
Fax: 508-647-7001
E-mail: info@mathworks.com
Web: www.mathworks.com

Note to Readers

Note to Readers on the Structure of the Book

This book is organized into *seven* chapters and *six* appendices, with each chapter typically having the following three sections: *brief theory, computer laboratory*, and *hardware laboratory*. All seven chapters have theory and computer laboratory sections; however, Chapters 1, 2, 3, and 4 also have a hardware section. Generally, each chapter includes a brief theory section, followed by a MATLAB/Simulink simulation section, and, finally, a hardware section, which includes building practical communication circuits and testing these circuits using signal generators, digital oscilloscopes, and spectrum analyzers.

This three-pronged approach is aimed at taking students from theory to simulation to experiment in a very effective way. Additionally, instructors have the option of selecting only the computer laboratory, only the hardware laboratory, or both for their classes based on the availability of software or hardware.

- *Guidelines for instructors:* Please note that in each chapter, each of the three sections (theory, computer lab, and, where applicable, hardware lab) contain *exercises* for students. Each chapter typically has about *10 total exercises* each, and the instructor can assign any or all the exercises for the students.

- *Guidelines for students:* Please attempt all exercises systematically or as assigned by your instructor, after reviewing the theory material in each chapter. Clarify all doubts with the instructor before proceeding to the next section, since each section draws information from the previous material.

Author

B. **Preetham Kumar** received his BE (electronics and communications) in 1982, ME (communications systems) in 1984, both from the College of Engineering, Chennai, India, and PhD in Electrical Engineering from the Indian Institute of Technology (IIT), Chennai, India, in 1993. He worked as a researcher and lecturer in the RF & Microwave Laboratory, University of California, Davis, and also as a part-time faculty member in the Department of Electrical and Electronic Engineering, California State University, Sacramento (CSUS), from 1993 to 1999. He joined CSUS on a full-time basis in August 1999, where he is currently professor and graduate coordinator in the Department of Electrical and Electronic Engineering.

He has advised around 200 master's student projects and published more than 60 papers in peer-refereed journals and international conferences in the areas of antenna design, RF, and microwave circuits. He has also published a book, *Digital Signal Processing Laboratory*, with CRC Press (two editions) and has coauthored two book chapters on power dividers and microwave/RF multipliers for the *Wiley Encyclopedia of Electrical and Electronics Engineering*.

Dr. Kumar has worked on research projects funded by Lockheed Martin, Agilent Technologies, National Semiconductor, Intel, and Antenna Wireless Inc. Currently, he is heading an effort to expand the use of hyperthermia, or microwave heating, as an adjuvant tool with radiation and chemotherapy in the treatment of cancer. He is the recipient of multiple International Cancer Technology Transfer (ICRETT) fellowships by the Union for International Cancer Control (UICC) in Geneva, Switzerland, to initiate clinical trials involving hyperthermia and radiation in India.

Dr. Kumar has received several awards, including the Tau Beta Pi Outstanding Faculty Award in 1999, the Outstanding Teaching Award in 2000, and the Outstanding Scholar Award in 2005 from the College of Engineering and Computer Science, CSUS. He is a senior member of the Institute of Electrical and Electronics Engineers (IEEE) and currently serves as the vice-chairman of the IEEE Sacramento Valley section and chairman of the Sacramento chapter of the IEEE Communications Society. He was elected in 2007 as overseas member in the Executive Council, the Indian Association of Hyperthermia Oncology and Medicine.

1

Types of Electronic Communication Systems

From the beginning of time, communication has been a basic function of all living beings to reach out to one another, or to convey messages or information of some kind. The message or information could be very vital, for example, a telephone call between two heads of state on a *hotline* could possibly affect the outcome of history, or the message could be just a simple hi-five between two people in the same room, or even birds calling out to each other.

While the basic purpose of communication is to convey information, factors that have evolved and improved over the years are the length and breadth of the system. Early civilizations used carrier birds (such as pigeons and owls, like the adorable *Hedwig* from *Harry Potter*) or actual people, as "runners" to transport messages from place to place (recall the fabled run of the Greek soldier *Pheidippides*, from the Battle of Marathon to Athens). People even used *drums* and *smoke signals* to transmit messages across inaccessible terrains, or flags displaying a *semaphore* code to communicate between ships.

However, the birth of *electricity* marked a renaissance in communications technology, based on conversion of sounds and images to electrical signals, which could be transmitted almost instantly over long distances. Electricity led to the development of the *telegraph, telephone, television, satellite*, and *cellular* communications. Initially, telegraph and telephone systems used *wires* for transmitting signals from place to place; however, with the advent of *wireless signaling* by Marconi, literally the whole earth was connected. In wireless, information is carried by *electromagnetic waves* which travel at the speed of light in air or vacuum. The simple relation between the frequency and wavelength of an electromagnetic wave is given by the following equation:

$$c = f\lambda \qquad (1.1)$$

where
f is the frequency of the wave in Hz
λ is the wavelength in m
c is the velocity of light

Table 1.1 illustrates the different radio-frequency bands that are in use today, and their practical applications.

TABLE 1.1

Radio-Frequency Bands and Applications

Frequency Band	Frequency Range	Wavelength Range	Application
Very low frequency (VLF)	3–30 kHz	10–100 km	Navigation, wireless monitors
Low frequency (LF)	30–300 kHz	1–10 km	Navigation, long wave AM, RFID, amateur radio
Medium frequency (MF)	300–3000 kHz	100–1000 m	Medium wave AM, amateur radio
High frequency (HF)	3 to 30 MHz	10–100 m	Shortwave AM, RFID, RF diathermy, medical applications
Very high frequency (VHF)	30–300 MHz	1–10 m	FM, television
Ultrahigh frequency (UHF)	300–3000 MHz	10–100 cm	Television, Cell phones, GPS, Wi-Fi, microwave oven, medical applications
Super high frequency (SHF)	3–30 GHz	1–10 cm	Satellite, radar, remote sensing, wireless LAN
Extremely high frequency (EHF)	30–300 GHz	1–10 mm	Satellite, radio astronomy
Tremendously high frequency (THF)	300–3000 GHz	0.1–1 mm	Terahertz communication, imaging

Example

(a) By referring to the Internet, determine the years in which the following technologies were available for public use: *telegraph, telephone, television, satellite,* and *cellular.*

(b) Categorize the technologies listed in (a) as *wired* or *wireless* in older and current day technology.

Solution

(a)

Electric telegraph:	1838 (Paddington station to West Drayton, United Kingdom)
	1838 (Capitol in Washington to old Mt. Clare Depot in Baltimore)
Telephone:	1878 (First commercial North American telephone exchange in New Haven, Connecticut)
Television:	1941 (Commercial transmission begins by NBC and CBS, New York)
Satellite:	1976 (Commercial satellite TV begins by HBO, New York and TBS, Atlanta)
Cellular:	1979 (Commercial cellular covered by NTT, Japan)

(b)

Electric telegraph:	Wired
Telephone:	Wired
Television:	Wired (cable) or wireless (dish)
Satellite:	Wireless
Cellular:	Wireless

1.1 How the World Is Linked through Coaxial, Microwave, Satellite, Cable, and Cellular Technologies

Surprisingly, even a simple phone call between two people involves the use of more than one type of communications technology. Figure 1.1 illustrates a typical link between subscribers located in different geographical areas (see the smart phones around the two cell towers). The subscribers connect to the cell towers using *cellular links*, while the towers connect to each other using *microwave links* and finally, the towers connect to the core communication network using wideband *fiber-optic cable*. These individual links vary in structure, complexity, bandwidth, and reliability.

In addition to cellular, microwave, and land-based cable links, connections between continents is provided by undersea *submarine cables* or overhead *satellite networks*. Figure 1.2 shows submarine cable networks crossing the oceans of the world. This map is a continuously changing profile as more networks are added to the global system.

Similarly, Figure 1.3 shows the iridium satellite network, which is a large group of 66 satellites providing voice and data coverage to satellite phones, pagers, and integrated transceivers over Earth's entire surface.

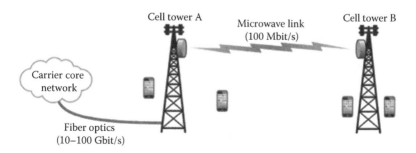

FIGURE 1.1
Communication network linking two phones.

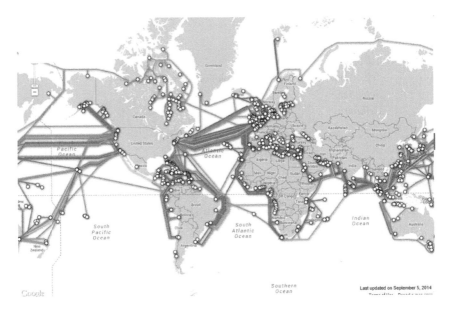

FIGURE 1.2
World submarine cable network. (Courtesy of TeleGeography, Washington, DC, www.submarinecablemap.com.)

FIGURE 1.3
Iridium satellite network. (Courtesy of AST Networks, Chandler, AZ, http://www.ast-systems.us.com/Networks/Iridium.aspx.)

Example

Draw an approximate link between a cell phone user in the United States and another cell phone user in Europe, using a combination of various technologies discussed earlier. Label the different sublinks and also the type of interface (e.g., Wireless, cable).

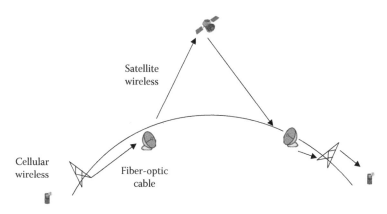

FIGURE 1.4
Connection path between two cellular users in different continents.

Solution

Figure 1.4 shows one solution to connect two cellular users in different continents.

1.2 Functional Layers in Modern Communication Systems

Engineers like to describe the global communication system as consisting of different layers, similar, for example, to the structure in an industrial environment: we have the *management* layer, which plans strategies to maintain a steady product output, and the *worker* layer, which actually manufactures the products. Likewise in the vast communications systems that link people around the world, we can define many layers, the most important being the *network layer* and *physical layer*. The *network layer* merges and controls different technologies such as microwave, satellite, cellular, and systems in different areas around the country and the world. The other important component, the *physical layer*, comprises the actual medium, which provides the communication and transmits the data between users, through either wireless or wired channels.

Example

Referring to the Internet, find at least three common mobile phone functions that are managed through the (a) *network layer* and (b) *physical layer* in modern communication systems.

Solution

(a) The important functions of the networking layer include the following:

(i) Logical addressing: Every device that communicates over a network has a digital logical address associated with it. For example, on the Internet, the *Internet Protocol* (*IP*) is the network layer protocol and every machine has an *IP address*.

(ii) Routing: Data or information is usually sent in digital systems as *packets*, and moving data from one network to other connected networks is probably the defining function of the network layer. Devices and software routines function at the network layer to process incoming packets from various sources, determine their final destination, and then figure out where they need to be sent to get them where they are supposed to go.

(iii) Error handling and diagnostics: Special protocols are used at the network layer to allow devices that are logically connected, or that are trying to route traffic, to exchange information about the status of hosts on the network or the devices themselves.

(b) The functions of the physical layer include the following:

(i) Hardware control and usage: The details of operation of cables, connectors, wireless radio transceivers, network interface cards, and other hardware devices are generally a function of the physical layer.

(ii) Encoding and signaling: The physical layer is responsible for various encoding and signaling functions that transform the data from bits that reside within a computer or other device into signals that can be sent over the network.

(iii) Data transmission and reception: After encoding the data appropriately, the physical layer actually transmits and receives the data, whether it comes through a wired or wireless channel.

(iv) Hardware network: The physical layer also serves to design hardware-related networks, such as *Local Area Networks* (*LANs*).

1.3 Path Loss in Communication Links

One of the main challenges in communication systems is the *loss of power* as signals travel over different media and terrain. For example, when a *cellular signal* travels through a city, it weakens as it gets reflected from obstacles such as buildings, cars, trees, and even street lights. To define this process of signal weakening from the transmitter to the receiver, we define the *path loss* (*PL*) in the following equation:

$$PL = \frac{\text{Transmitted power, W}}{\text{Received power, W}} \qquad (1.2)$$

Since PL is a ratio of two powers in Watts, it has no units. More practically, PL is given in logarithmic units (dB) as shown in the following equation:

$$PL, \text{dB} = \text{Transmitted power, dBm} - \text{Received power, dBm} \qquad (1.3)$$

where the logarithmic unit of power, P, $\text{dBm} = 10 \log_{10}(P, \text{mW})$. For example, 1 W of power, or 1000 mW, would correspond to 30 dBm. Hence, it is important to understand that dBm is an absolute unit of power, while dB represents a ratio such as gain or loss.

Example

(a) Determine the unit PL (dB/km) of a 60 km undersea cable, if the transmitter power is 10 W and the received power is 8 W.
(b) Determine the received power if the cable is extended to 80 km.

Solution

(a) PL, dB = Transmitted power, dBm − Received power, dBm
= 10 $\log_{10}(10,000)$ − 10 $\log_{10}(8,000)$, after converted W to mW.
= 40 − 39.03 = 0.969 dB
Hence, unit PL = Cable PL/Cable length
= 0.969/60 = 0.016 dB/km
(b) PL for 80 km cable = Unit PL × Length of cable
= 0.016 × 80 = 1.292 dB
Hence, from Equation 1.3,
Received power, dBm = Transmitted power, dBm − PL, dB
= 40 − 1.292 = 38.708 dBm

1.4 Introduction to MATLAB®/Simulink®

Programming software that is utilized in communication systems modeling and applications can be categorized into the following two categories:

- Simulation software: These software, such as MATLAB® and Simulink®, are utilized to model communication systems, and hence are very valuable tools to design practical systems. While MATLAB requires programs to be written, Simulink is a graphical tool, which has built-in system blocks.

- Software controlling hardware: This software is required to run communication systems hardware such as Digital Signal Processors (DSPs). Examples of this kind of software are assembly language and C.

1.4.1 MATLAB® Basics

Please try out each of the commands given in the following and familiarize yourself with the different types of MATLAB commands and formats.

1.4.1.1 System Operating Commands

MATLAB can be opened by clicking on the computer desktop MATLAB "icon". The MATLAB prompt is >>, which indicates that commands can be started, either line by line, or by running a program. A complete program, consisting of a set of commands, can be saved in a MATLAB file for repeated use as follows:

- Open a file in any text editor (either in MATLAB or otherwise), and write the program.
- After writing the program, exit saving as a *filename.m* file.
- To execute the program, either run the file from the text editor or type the filename after the prompt:

  ```
  >> filename
  ```

The program will run, and the results and error messages, if any, will be displayed on the screen. Plots will appear on a new screen.

1.4.1.2 Numbers

Generation of numbers

Example: Generate the real numbers $z1 = 3$, $z2 = 4$.
```
>> z1 = 3
>> z2 = 4
```

Example: Generate the complex numbers $z1 = 3 + j4$, $z2 = 4 + j5$.
```
>> z1 = 3+j*4
>> z2 = 4+j*5
```
Note: The symbol i can be used instead of j to represent $\sqrt{-1}$.

Example: Find the magnitude and phase of the complex number $3 + j4$.
```
>> z = 3+j*4
>> zm = abs(z)        ; gives the magnitude of z
>> zp = angle(z)      ; gives the phase of z in radians
```

Addition or subtraction of numbers (real or complex)

```
>> z = z1 + z2        ; addition
>> z = z1 - z2        ; subtraction
```

Multiplication or division of numbers (real or complex)

```
>> z = z1*z2          ; multiplication
>> z = z1/z2          ; division
```

1.4.1.3 Vectors and Matrices

Generation of vectors

Example: Generate the vectors $x = [1\ 3\ 5]$ and $y = [\ 2\ 0\ 4\ 5\ 6]$.

```
>> x = [1 3 5]        ; generates the vector of length 3
>> y = [2 0 4 5 6]    ; generates the vector of length 5
```

Addition or subtraction of vectors x *and* y *of* same length

```
>> z = x + y          ; addition
>> z = x - y          ; subtraction
```

Multiplication or division of vectors x *and* y *of* same length

```
>> z = x. * y         ; multiplication
>> z = x. / y         ; division
```

Note: The dot after x is necessary, since x is a vector and not a number.

1.4.1.4 Creating One-Dimensional and Two-Dimensional Spaces Using MATLAB®

The command:

```
>> x = linspace(x1,x2,N)   ; generates N points between x1
                             and x2
                           ; stores it in the vector x
```

The following three commands generate a two-dimensional mesh:

```
>> x = linspace(x1,x2,N1)  ; generates N1 points between
                             x1 and x2
                           ; stores it in the vector x
>> y = linspace(y1,y2,N2)  ; generates N2 points between
                             y1 and y2
                           ; stores it in the vector y
>> [X,Y] = meshgrid(x,y)   ; generates the two-dimensional
                             matrix [X,Y]
```

Example

Plot the following continuous-time signals using MATLAB. If you decide that the signal is periodic, select a range for time t that will cover at least three time periods.

(a) $x(t) = 5 \cos[\, 2\pi(15)t + 0.25\pi]$
(b) $x(t) = 5 \cos[\, 2\pi(15)t - 0.5\pi] + 5 \cos[\, 2\pi(10)t + 0.3\pi]$

Solution

```
% MATLAB program
  clear
  t=0:0.001:0.5;
  x=5*cos(2*pi*15*t+0.25*pi);
  subplot(2,1,1);
  plot(t,x)
  xlabel('t,sec');
  ylabel('x(t)');
  title('Problem a');

  t=0:0.001:0.8;
  x=5*cos(2*pi*15*t - 0.5*pi) + 5*cos(2*pi*10*t + 0.3*pi);
  subplot(2,1,2);
  plot(t,x)
  xlabel('t,sec');
  ylabel('x(t)');
  title('Problem b');
```

The plots are shown in Figure 1.5 as Problems (a) and (b), respectively.

1.4.1.5 Programming with Vectors

Programs involving vectors can be written using either *For Loops* or *vector* commands. Since MATLAB is basically a vector-based program, it is often more efficient to write programs using *vector* commands. However, *For Loops* give a clearer understanding of the program, especially for the beginner:

Example: Sum the following series:

$S = 1 + 3 + 5 \cdots \cdots 99.$

For loop approach

```
>> S = 0.0                  ; initializes the sum to zero
>> for i = 1 : 2 : 99
    S = S + i
    end
>> S                        ; gives the value of the sum
```

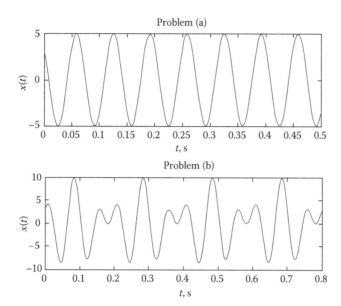

FIGURE 1.5
MATLAB® plots for continuous-time signals.

Vector approach

```
>> i = 1: 2 : 99;    ; creates the vector i
>> S = sum ( i );    ; obtains the sum S
```

Note: The use of a semicolon after any MATLAB command causes that command output not to display on the screen. If the semicolon is not used, then each command will be displayed on the screen, which might be time consuming when large loop statements are used.

Example: Generate the discrete-time signal $y(n) = n \sin(\pi n/2)$ in the interval $0 \le n \le 10$.

For Loop approach

```
>> for n = 1:1: 11
   n1(n) = n - 1
   y(n) = n1(n) * sin(pi*n1(n)/2)
   end
>> y                 ; gives the vector y
>> stem(n1,y)        ; plots the signal y vs. n with impulses
```

Vector approach

```
>> n = 0 : 10;            ; creates the vector n
>> y = n.*sin(pi*n/2);    ; obtains the vector y
>> stem(n,y)             ; plots the signal y vs. n with
                           impulses
```

Example

Define the discrete-time signal $x(n) = n$, in a vector in the range $0 \leq n \leq 10$ and plot the signal.

Solution

```
>> n = 0:10      ; defines the vector n of length 11
>> x = n         ; defines the signal x(n)
>> stem(n,x)     ; plots the discrete signal x(n)
```

Example

Plot the following discrete-time signals using MATLAB. If you decide that the signal is periodic, select a range for time t that will cover at least three time periods.

(a) $x(n) = 0.5^{|n|}$, in the range $-5 \leq n \leq 5$
(b) $x(n) = 5 \cos(\omega n + 0.3\pi)$, $\omega = 1.5\pi$.

Solution

```
% MATLAB program
clear
n=-5:5;
x=0.5.^abs(n);
subplot(2,1,1);
stem(n,x)
xlabel('n');
ylabel('x(n)');
title('Problem a');

clear
n=-10:10;
x=5*cos(1.5*pi*n+0.3*pi);
subplot(2,1,2);
stem(n,x)
xlabel('n');
ylabel('x(n)');
title('Problem b');
```

The plots are shown in Figure 1.6 as Problems (a) and (b), respectively.

1.4.2 Simulink® Basics

After logging into MATLAB, you will receive the prompt >>. In order to open up *Simulink*, type in the following:

```
>> Simulink
```

Alternately, click on the Simulink icon in the MATLAB Command window, or in newer editions of the software, use the command *Open New Model*.

1.4.2.1 General Simulink® Operations

Two windows will open up: the *model window* and the *library window*. The *model window* is the space utilized for creating your simulation model.

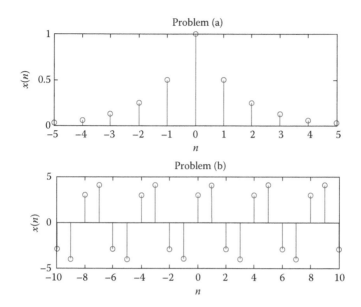

FIGURE 1.6
MATLAB® plots for discrete-time signals.

In order to create the model of the system, components will have to be selected and dragged from the library using the computer mouse, and inserted into the model window. If you browse the library window, the following sections will be seen. Each section can be accessed by clicking on it.

- Sources—This section consists of different signal sources such as sinusoidal, triangular, pulse, random, or files containing audio or video signals.
- Sinks—This section consists of measuring instruments such as scopes and displays.
- Linear—This section consists of linear components that perform operations like summing, integration, and product.
- Nonlinear—Nonlinear operations.
- Connections—Multiplexers, demultiplexers.
- Toolboxes—These specify different areas of Simulink, for example:
  - Communications
  - DSP
  - Neural nets
  - Simscape (power systems)
  - Control systems

1.4.2.2 Editing, Running, and Saving Simulink® Files

The complete system is created in the model window by utilizing components from the various available libraries. Once a complete model is created, *save* the model into a file. Click on *simulation* and select *run*, or in newer models, click on the *green arrow*. The simulation will run, and the output data can be read by clicking on the appropriate *sinks*, such as oscilloscopes or displays. Save the output plots also into files. The model and output files can be printed out from the files.

1.4.2.3 Demo Files

Try out the demo files, both in the *main library* window, and in the *Toolboxes* window. There are several illustrative demonstration files in the areas of *signal processing, image processing,* and *communications.*

1.5 Introduction to Equipment Used in Communication Systems

Hardware equipment utilized in communication systems applications can be classified into three main categories:

1.5.1 Sources

Sources generate signals that vary in shape, amplitude, frequency, and phase. One popular signal source is the *Keysight 33600A waveform function generator.* Please see Appendix A for the manufacturer's specification details, and other operating suggestions.

1.5.2 Measurement Devices

Measuring devices are utilized to accurately graph input signals in two domains: *time* and *frequency.* The *Keysight DS090254A digital storage oscilloscope* measures the amplitude waveform of signals as a function of time, whereas the *Keysight N9000A signal analyzer* measures the spectrum of the input signal as a function of frequency. Please see Appendices B and D for manufacturer's sheet and details of these two measuring equipment.

The *Keysight 35670A dynamic signal analyzer* is multipurpose equipment that can measure signals in both time and frequency domains, in addition to frequency response of devices. These analyzers are more advanced equipment that can also generate regular signals, as well as random noise. Please see Appendix C for manufacturer's details.

1.5.3 Radio-Frequency Integrated Circuits

Radio-frequency integrated circuits (RFICs) are the most widely used components in many communications systems including cell phones. They are essentially Systems of Chip (SoC) which can perform many important radio functions such as filtering, modulation, and Fast Fourier Transform (FFT). RFICs are manufactured by companies like Texas Instruments, Motorola, and Intel; please read Appendix E for details of some typical RFICs.

Problem Solving

P1.1 Communication systems primarily cover the transmission of information using the following technologies:

(a) Coaxial cable

(b) Microwave link

(c) Satellite link

(d) Submarine cable

(e) Cellular link

(f) Fiber-optic link

Research the Internet to get information on the approximate *frequency ranges* of these six communication system technologies.

P1.2 Current wireless technology has a wide terminology relating to different systems:

(a) Wi-Fi

(b) WiMax

(c) Bluetooth

(d) Zigbee

(e) 3G

(f) 4G

(g) 5G

Research the Internet to get information on the approximate *frequency ranges* of these wireless systems in the United States, Europe, and Asia.

P1.3 Referring to Figure 1.4, which shows the communication path between two users in different continents, please find the system parameters here:

(a) Transmitter power (from cell phone in first continent) = 1 W
(b) Cellular PL (between user and cell tower) = 20 dB
(c) Cellular tower amplification = 30 dB
(d) PL between tower and dish antenna = 2 dB
(e) PL between dish antenna and satellite = 60 dB

Assuming same parameters on the receiving side, calculate the following:

(a) Net PL between the two users in dB
(b) Received power (at cell phone in second continent), in Watts and dBm

Also, comment on the different levels of PL for cellular, fiber-optic cable, and satellite links.

P1.4 Approximately sketch the following signals in the range $0 \leq t, s \leq 10$:

(a) $x(t) = u(t) - u(t - 3)$
(b) $x(t) = u(3 - t)$
(c) $x(t) = 0.5^t [u(t) - u(t - 5)]$

In all these examples, $u(t)$ is defined as the *unit step* function, as follows:

$$u(t) = \begin{cases} 1, & t \geq 0 \\ 0, & t < 0 \end{cases}$$

P1.5 Determine if each of the following signals is *periodic* or *nonperiodic*. If a signal is periodic, specify its fundamental period:

(a) $x(t) = e^{j\pi t}$
(b) $x(t) = 0.7^t$
(c) $x(t) = \cos(\pi t/2) \cos(\pi t/4)$

P1.6 A digital audio signal can be represented by a one-dimensional vector $x(n)$; however, a digital video signal requires a two-dimensional matrix $x(m, n)$. Describe two reasons why it is more difficult to process and transmit video signals as compared to audio signals.

Computer Laboratory

C1.1 Write a MATLAB program to sketch the following continuous-time signals in the time range of $-10 \le t \le 10$ s. Please label all the graph axes clearly. If the sequence is complex, plot the magnitude and angle separately.

(a) $x(t) = 2t^2 - 3t^3$

(b) $x(t) = \sin(\pi t/3)$

(c) $x(t) = 0.5^t \, e^{j\pi t/2}$

C1.2 Write a MATLAB program to sketch the following discrete-time signals in the time range of $-10 \le n \le 10$. Please label all the graph axes clearly. If the sequence is complex, plot the magnitude and angle separately.

(a) $x(n) = 2n^2 - 3n^3$

(b) $x(n) = \sin(\pi n/3)$

(c) $x(n) = 0.5^n \, e^{j\pi n/2}$

C1.3 There are two main forms of vector or matrix multiplication. In MATLAB, if two vectors a and b are given, then the two possible MATLAB multiplication commands are: y = a * b and y = a.*b.

(a)   Comment on the differences between these two commands, and clearly state the outputs of these two operations.

(b)   Write a brief MATLAB program to evaluate these two commands for the case of a = [1 2 3] and b = [4 5 6]. Are both the operations a * b and a.*b possible? If not, what change in the syntax would make the operation possible?

C1.4 Create Simulink models for the continuous-time system shown in Figure 1.7.
Before starting any simulation, please select the *simulation* button from the model window, and then select the *parameter* button. Modify the *start* time and *stop* time of the simulation to complete at least two periods of the signal source.

(a)   Run the simulation for sinusoidal signal, $x(t)$, with amplitude $A = 5$ V and frequency $\omega = 10$ rad/s. The signal $n(t)$ is a pseudorandom noise with maximum amplitude of 0.5 V. Obtain a

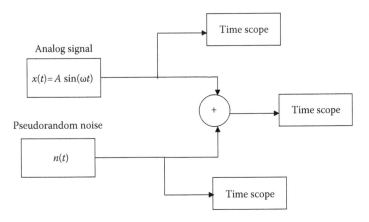

FIGURE 1.7
Simulink® model for continuous-time system.

printout of the combined output signal $y(t)$ on the time scope, and familiarize yourself with the settings.

(b) Change sinusoidal signal amplitude (2 V, 10 V), and frequency (20 rad/s, 50 rad/s), and observe the output on the time scope.

C1.5 Create Simulink models for the discrete-time system shown in Figure 1.8.
Please note that in all discrete-time simulation blocks, the appropriate sampling time, T, s should be specified.

(a) Obtain a plot of the output signal on the time scope, for an input signal of $x(t) = 3 \cos(2\pi t/5)$, in the time range $0 \le t \le 5$ s, when the signal is sampled at a time interval of $T = 0.5$ s.

(b) Change the input signal amplitude to 6 V, and input signal frequency to twice its original value, and observe on the time scope. Obtain a plot of the output signal. Comment on the differences between output signals obtained in both parts of this exercise.

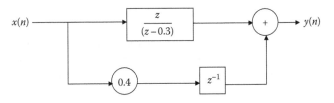

FIGURE 1.8
Simulink® model for discrete-time system.

Hardware Laboratory

H1.1 In this experiment, basic *time* and *frequency* measurements will be performed using the *oscilloscope* and *signal analyzer*.

(a) Connect the equipment together as shown in the schematic in Figure 1.9. Use BNC cables and a BNC Tee to connect the circuit. Please ensure that the Keysight 33250A Synthesized sweep generator POWER button is in the *OFF* position.

(b) Set the sweep generator to output a sinusoidal signal, with an amplitude of 5 V and frequency f = 2 MHz. Observe the time-domain signal output on the oscilloscope, and note down the measured amplitude and frequency of the sinusoidal signal.

(c) Set the signal analyzer to a START frequency of 1.0 MHz and a STOP frequency of 3.0 MHz. Observe the frequency-domain output on the signal analyzer, and note down the measured amplitude and frequency of the sinusoidal signal. Use the MARKERS in the signal analyzer to PEAK SEARCH mode to track the peak value in the signal spectrum.

(d) Comment on the differences, if any, among the set sweep generator frequency, oscilloscope output frequency, and signal analyzer output frequency.

H1.2 In this experiment, time and frequency measurements will be performed using the Keysight 35670A dynamic signal analyzer. The dynamic signal analyzer is a very versatile low-frequency equipment that can analyze and manipulate signals in the frequency range of 0–50 kHz.

(a) The dynamic signal analyzer has one output port called *source*, and two input ports called *channel* 1 and *channel* 2. The output of the *source* port is controlled by the *source* key on the top

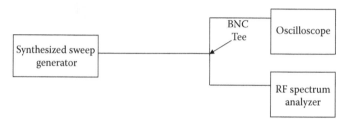

FIGURE 1.9
Measurement setup for time and frequency measurements.

section of the signal analyzer. The *source* can generate several kinds of sources including single frequency sinusoidal, swept frequency sinusoidal, and random noise sources.

(b) Connect the *source* output to *channel* 1 input with a BNC cable. Select the *source* key, and select *sinusoidal* source with a frequency of 10 kHz and amplitude of 5 V. Select the *measurement* key, and alternate between *time* and *frequency* settings to observe the signal in both domains. Time and frequency plots of the signal can be viewed simultaneously by using the *dual channel* display mode.

(c) Repeat the previous step of this experiment with a *random* signal source having a peak amplitude of 1 V. Observe the random signal in both time and frequency domains.

H1.3 In this experiment, frequency response measurements will be performed using the Keysight 35670A dynamic signal analyzer.

(a) The dynamic signal analyzer can measure the frequency response of a passive device, for example, an electrical filter in the frequency range of 0–50 kHz. In this experiment, we will determine the frequency response of a low pass filter, having a cutoff frequency of 10 kHz.

(b) The filter circuit is shown in Figure 1.10, with port 1 as the *input port* and port 2 as the *output port*. Since the circuit is quite simple, it can be put together even on a breadboard for testing. However, if a circuit is soldered together, that would be more ideal.

(c) Connect the filter circuit to the Keysight 35670 signal analyzer as shown in Figure 1.11.

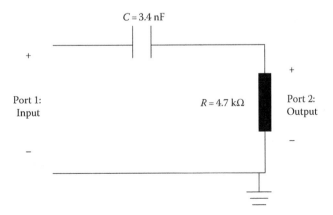

FIGURE 1.10
Basic RC filter circuit.

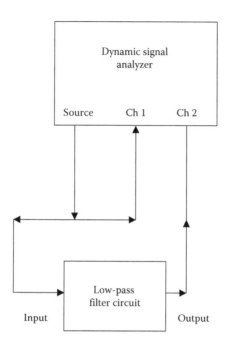

FIGURE 1.11
Measurement setup for frequency response measurement.

(d) The source output of the signal analyzer is simultaneously connected to the input port of the filter circuit and also to channel 1 of the signal analyzer, while the output port of the filter circuit is connected to channel 2 of the signal analyzer.

(e) It is important to set the signal analyzer settings appropriately to obtain the frequency response of the filter circuit. Select the *frequency* key on the signal analyzer, and set the *start* and *stop* frequencies to 0 Hz and 20 kHz, respectively. Select the *measure* key, and set the measurement to 2 *channel* measurements, and then select *frequency response* setting. Select the *source* key, and set the source to *chirp* signal, which will generate a swept frequency signal for 0–20 kHz. Set the *amplitude* of the chirp signal to 1 V.

(f) Select the *source* key, and set it to the *on* position. Finally select the *scale* key, and set it at *autoscale*. The frequency response of the filter should now appear on the screen. Obtain a record of this response, and note down key parameters like filter gain and 3 dB cutoff frequency.

2

Time/Frequency Analysis of Communication Signals and Systems

In Chapter 1, we introduced various types of communication systems such as satellite, cellular, submarine, and fiber-optic cable networks. Systems like coaxial, submarine cable, and fiber optics use *wired* channels for transmitting signals, while cellular and satellite systems use *wireless* channels. The signals going through these wired or wireless transmission channels are quite complex, and it is important to understand their *time* and *frequency content* to design efficient transmission and receiver systems. For example, if we know that a particular signal is centered at 92.5 MHz (the frequency of a popular FM station), then we can use a *band-pass filter* to extract this signal.

2.1 Concept of Carrier in Communication Systems

What kinds of electrical signals flow through communication systems? To answer this question, we need to recognize that communication systems are essentially *transporters* that carry information across the world in the form of telephone calls, e-mails, texts, Internet, radio, and television, which are nowadays a part of our daily life. These transport systems can be compared to trains or planes that move people from place to place; just as a modern commercial plane is a strong and reliable transport, similarly there is need for a strong, stable signal to carry the information. The latter signal is called as the *carrier*, and is just a simple sinusoid given by the following equation:

$$x(t) = A \cos(\omega_c t + \theta) \tag{2.1}$$

where
A is carrier amplitude, V
ω_c is the carrier frequency radians/s
θ is the carrier phase (rad)

The carrier frequency is fc (Hz) $= \omega_c/2\pi$, and varies with application, from kHz to GHz, as shown in Table 2.1.

In addition to frequency, important properties of the carrier signal include its *power* and *bandwidth*, which controls data speed, including upload and download speeds.

TABLE 2.1

Communication Systems and Frequency Bands

Application	Carrier Frequency
AM radio	Long wave (LF): 153–279 kHz
	Medium wave (MF) 531–1611 kHz
	Short wave (HF) 2.3–26.1 MHz
FM radio	76.0–108.0 MHz
Satellite	1–75 GHz
Line of sight (LOS) microwave links	1–25 GHz
Cellular	800 MHz–2.5 GHz

- *Power* in any time signal $x(t)$ can be calculated as follows:

$$P_x = \lim_{T=\infty} \frac{1}{T} \int_{-T/2}^{T/2} |x(t)|^2 dt \qquad (2.2)$$

The unit of power is Watts (W), or dBm, and the symbol $|\ |$ denotes the absolute value or magnitude, since the signal may be complex. For periodic signals, the integration is carried out over one period of the signal:

$$P_x = \frac{1}{T} \int_{0}^{T} |x(t)|^2 dt \qquad (2.3)$$

The time period of a periodic signal is related to its frequency as $T = 2\pi/\omega_0$.

- The *bandwidth* of a signal is an important property that governs the performance of communication systems. Simply put, the bandwidth of a signal is the *difference between its maximum frequency and minimum frequency content*; for example, if a signal has frequency content from 1 to 2.4 MHz, then its bandwidth is $2.4 - 1 = 1.4$ MHz. Most practical signals that we use commonly are *bandlimited*, which means that the bandwidth has a finite value; for example, audio signals have a maximum bandwidth of 20 kHz, while video signals have a larger bandwidth of 5 MHz.

Interestingly, the bandwidth of communication systems has steadily increased over the decades, to provide more services to users such as video and high-speed Internet networks. In fact, as will be seen later, *Shannon's theorem* states that the transmission speed of communication systems is directly proportional to the bandwidth of the channel, that is, the higher the bandwidth, the faster is the Internet!

Example

Find the power in a carrier signal $x(t) = 5 \cos(200\pi t + 45°)$, using both Watts and dBm units.

Solution

As defined in Equation 2.3, for a periodic signal:

$$P_x = \frac{1}{T} \int_0^T |x(t)|^2 dt$$

The period of the signal $T = 2\pi/\omega_0 = 2\pi/200\pi = 0.01$ s. Hence,

$$P_x = \frac{1}{T} \int_0^T |x(t)|^2 dt = \frac{1}{0.01} \int_0^{0.01} |5\cos(200\pi t + 45°)|^2 dt$$

$$= \frac{1}{0.01} \int_0^{0.01} 25\cos^2(200\pi t + 45°)\, dt$$

$$= \frac{1}{0.01} \int_0^{0.01} 25 \left[\frac{1 + \cos(400\pi t + 90°)}{2} \right] dt$$

$$= \frac{25}{0.02} = 1250 \text{ W}$$

Converting to dBm

$$P_x(\text{dBm}) = 10 \log_{10} P_x(\text{mW})$$

$$= 10 \log_{10} P_x(1,250,000) = 60.97 \text{ dBm}$$

2.2 Signal Spectrum and the Fourier Transform

The carrier signal, $x(t)$, as described in Equation 2.1 and shown in Figure 2.1a, has a single frequency $\omega = \omega_c$; hence in the frequency domain, it would be represented as shown in Figure 2.1b. However, *how does the frequency component look the way it is*? Why are there two peaks at ω_c and $-\omega_c$ (rad/s), both with amplitude $A/2$? The relation between the time and frequency representations is governed by the time-honored Fourier integral:

$$X(\omega) = \int_{-\infty}^{\infty} x(t)\, e^{-j\omega t} dt \qquad (2.4)$$

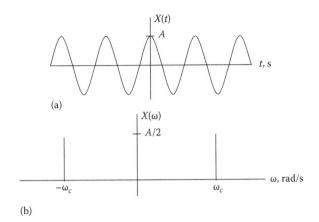

(a)

(b)

FIGURE 2.1
(a) Time and (b) frequency representations of a carrier signal.

Likewise, the inverse equation to convert from the frequency to time domain is given by the following equation:

$$x(t) = \frac{1}{2\pi} \int_{-\infty}^{\infty} X(\omega)\, e^{j\omega t} d\omega \qquad (2.5)$$

2.2.1 Important Facts about the Fourier Transform

2.2.1.1 Continuous and Discrete Spectrum

Generally, the spectrum of a signal $x(t)$ is *continuous*; however, periodic signals have a *discrete* spectrum. In general, a periodic signal with a period of T, s has a frequency of $f_0 = 1/T$ Hz or $\omega_0 = 2\pi/T$ rad/s. Then, the Fourier spectrum would be discrete, occurring at $0, f_0, 2f_0, 3f_0, \ldots, \infty$ Hz, or $0, \omega_0, 2\omega_0, 3\omega_0, \ldots, \infty$ rad/s. The component at 0 Hz is called as the *dc term*; the component at f_0 Hz is called as the *fundamental*; the component at $2f_0$ Hz is called as the *first harmonic*, the component at $3f_0$ Hz is called as the *second harmonic* and so on.

For a periodic signal, the Fourier transform pair, given in Equations 2.4 and 2.5, reduces to the following equations, respectively; Equation 2.6 obtains the discrete Fourier coefficients, X_n, $-\infty \le n \le \infty$; at the frequency $n\omega_0$ and Equation 2.7 expands the signal as a discrete sum. The latter expansion is termed as the *Fourier series*.

$$X_n = \frac{1}{T} \int_{0}^{T} x(t)\, e^{-jn\omega_0 t} dt \qquad (2.6)$$

$$x(t) = \sum_{n=-\infty}^{\infty} X_n e^{jn\omega_0 t} \qquad (2.7)$$

Example

Consider the periodic signal shown in Figure 2.2:

(a) Determine the signal frequency in Hz and rad/s, and the power of the signal.

(b) Identify the frequencies corresponding to the dc term, and fundamental and first *three* harmonics.

(c) Determine the numerical values of the Fourier coefficients, X_n, for $n = 0, \pm1, \pm2,$ and ±3.

Solution

(a) By examining the signal pattern in Figure 2.2, the signal period is found to be 3 ms; hence, the frequency is

$f_0 = 1/3$ ms $= 333.33$ Hz

and

$\omega_0 = 2\pi \times 333.33 = 2094.40$ rad/s

As defined in Equations 2.4 and 2.5, for a periodic signal:

$$P_x = \frac{1}{T}\int_0^T |x(t)|^2 dt$$

The period of the signal $T = 3$ ms $= 0.003$ s.

Hence,

$$P_x = \frac{1}{T}\int_0^T |x(t)|^2 dt = \frac{1}{0.003}\int_0^{0.001} |5^2| dt$$

$$= \frac{1}{0.003}\int_0^{0.001} 25\, dt$$

$$= \frac{25 \times 0.001}{0.003} = 8.33 \text{ W}$$

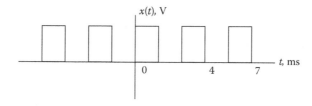

FIGURE 2.2
Periodic time signal.

TABLE 2.2

Amplitude and Phase of Fourier Coefficients

N	$|X_n|$	$/X_n$ degrees
-3	0.0000	180.0000
-2	0.6892	120.0000
-1	1.3783	60.0000
0	1.6700	0
1	1.3783	60.0000
2	0.6892	120.0000
3	0.0000	180.0000

(b) The dc frequency is 0 Hz; the fundamental frequency is 333.3 Hz and the first three harmonic frequencies are 666.6, 999.9, and 1333.2 Hz, respectively.

(c) Using Equation 2.6, Fourier coefficients are given as

$$X_n = \frac{1}{T} \int_0^T x(t)\, e^{-jn\omega_0 t}\, dt$$

$$= \frac{1}{0.003} \int_0^{0.001} 5\, e^{-jn(2\pi/0.003)t}\, dt$$

$$= \frac{5}{0.003} \left. \frac{e^{-jn(2\pi/0.003)t}}{-jn2\pi/0.003} \right|_0^{0.001}$$

$$= 5 \frac{\left[e^{-jn(2\pi/0.003)0.001} - 1 \right]}{-jn2\pi}$$

Calculating the values of the coefficients for different values of n, we obtain Table 2.2.

2.2.2 Power and Energy Relations in the Fourier Domain

The key point to understand the Fourier spectrum is that time and frequency are two *aspects* of the same signal; it is like looking at two sides of the same coin, they look different, but represent the same object. As an illustration, the time and frequency representations, $x(t)$ and $X(\omega)$ respectively, conserve *energy* in both domains, according to *Parseval's theorem*, which can be written as:

$$\text{Signal energy } E_x = \int_{-\infty}^{\infty} |x(t)|^2\, dt = \frac{1}{2\pi} \int_{-\infty}^{\infty} |X(\omega)|^2\, dt \tag{2.8}$$

Likewise, the signal power, for a *periodic signal* with period *T*, can be represented in both time and frequency domains as follows:

$$\text{Signal power } P_x = \frac{1}{T}\int_0^T |x(t)|^2\, dt = \sum_{n=-\infty}^{\infty} |X_n|^2 \tag{2.9}$$

Example

In the earlier example, calculate the power in the signal using the Fourier series formula in Equation 2.9, using the terms $n = -3$ to 3, and compare it with the exact power obtained earlier using the time domain approach.

Solution

Using Equation 2.9, the signal power is

$$P_x = \sum_{n=-\infty}^{\infty} |X_n|^2 \sim \sum_{n=-3}^{3} |X_n|^2$$

Adding the terms, we get

$$P_x \sim 0.0^2 + 0.6892^2 + 1.3783^2 + 1.6700^2 + 1.3783^2 + 0.6892^2 + 0.0^2 = 7.5383 \text{ W}$$

When comparing, this approximate value of the signal power is slightly different from the exact power of 8.33 W that we calculated earlier.

2.3 Important Communication Signals and Their Frequency Spectra

Some important time functions and their frequency spectrum, along with some practical applications, are listed in Figure 2.3a through d:

- *Impulse function*: $A\delta(t)$
  The practical implementation of the delta function is in *white noise*, which affects communication channels and has a relatively uniform distribution over the entire frequency range. The word "white" implies that the spectrum of the noise is uniform across the spectrum, similar to white color, which is a combination of all the colors of the spectrum.

- *Pulse function*: $A\Pi(t/T)$
  The practical implementation of the pulse function is in *digital communications*, where it represents the voltage waveform of the digital states 0 and 1. Looking at the spectrum of the pulse function, which

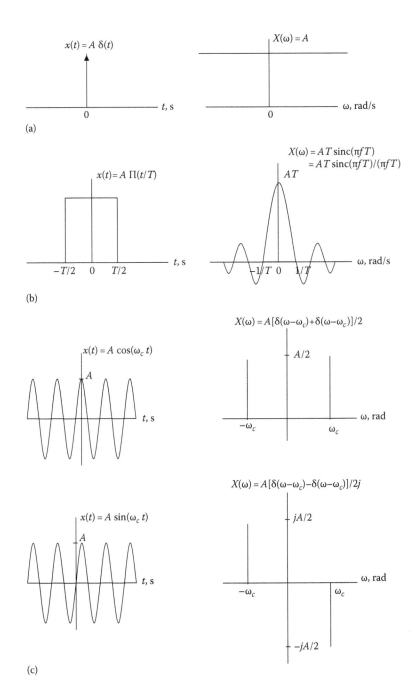

FIGURE 2.3
Time and frequency patterns of (a) impulse function, (b) pulse function, (c) sine/cosine functions.   (*Continued*)

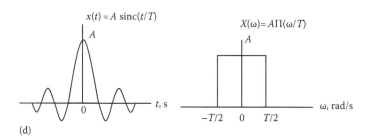

FIGURE 2.3 (*Continued*)
Time and frequency patterns of (d) pulse function.

extends infinitely or has infinite bandwidth, methods were evolved to *shape* the pulse and obtain a more bandlimited spectrum.

- *Sine/cosine functions: A* $\cos(\omega_c t)/A \sin(\omega_c t)$
  The practical implementation of the sine/cosine function is in *communications systems*, where it represents the fundamental carrier signal that carries information and data. The sine/cosine function is also important in *power systems*, where it is the primary waveform that carries power across transmission lines, and ultimately provides energy for residential and industrial users.

- *Sinc function: A* $\text{sinc}(t/T)$
  The practical implementation of the time sinc function is in *digital communications*, where it represents the *shaped pulse function*, to minimize interference between adjacent symbols. This pulse choice to represent a digital 0 or 1 is smoother and more rounded than a rectangular pulse function (the function discussed in the earlier section), which has increased bandwidth on account of its sharp edges.

  The last two functions discussed also show the *duality* nature of the Fourier transform, namely, a time rectangular pulse yields a sinc function as its spectrum, and inversely, a time sinc function yields a rectangular pulse as its spectrum!

Example

(a) Find the average power in the signal pulse signal, shown in Figure 2.4.
(b) Determine the approximate bandwidth of the signal, by truncating the Fourier spectrum of the signal at its first zero.

Solution

(a) From Equation 2.3, the power in a signal is given by

$$P_x = \lim_{T=\infty} \frac{1}{T} \int_{=T/2}^{T/2} |x(t)|^2 \, dt$$

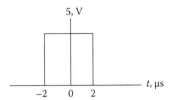

FIGURE 2.4
Pulse signal.

The signal given in the problem is not periodic; hence, its power is obtained by averaging its energy over the signal duration, as follows:

$$P_x = \frac{1}{T} \int_{-T.2}^{T/2} |x(t)|^2 \, dt$$

$$= \frac{1}{4} \int_{-2}^{2} 5^2 \, dt$$

$$= 25 \text{ W}$$

(b) From the transform given in Section 2.3, we can write the Fourier pair as

$$A\Pi\left(\frac{t}{T}\right) \leftrightarrow AT \sin c(\pi f T)$$

or, for this example: $5\Pi(t/4) \leftrightarrow 20 \sin c(4\pi f)$, which is sketched in Figure 2.5. From the condition given in the problem, and cutting off the spectrum at the first zero, bandwidth of $x(t)$ is $1/(4 \text{ μs}) = 0.25 \times 10^6$ Hz or 250 kHz.

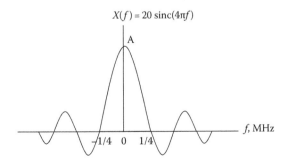

FIGURE 2.5
Sinc function.

2.4 Frequency Analysis of Communication Systems

This section deals with the analysis of signal transmission through communication systems, which could be *linear* or *nonlinear*. Linear systems are governed by the law of superposition, which basically states that the addition of two signals at the input of the system would yield the sum of two signals at the output, which correspond to the outputs when the two input signals are applied separately. Examples of linear systems include circuits built with resistors, capacitors, and inductors, while examples of nonlinear systems include circuits built with transistors.

2.4.1 Linear Systems

Figure 2.6 shows a linear time invariant (LTI) system, with input $x(t)$ and output $y(t)$, respectively. The system is represented by the impulse response, $h(t)$, which is defined as output of the system to the impulse input, $\delta(t)$. The input–output relation is given by the following *convolution* integral:

$$y(t) = x(t) * h(t)$$

$$= \int_{-\infty}^{\infty} x(p)\, h(t-p)\, dp \tag{2.10}$$

The relation between the input and output is much simpler in the frequency domain:

$$Y(\omega) = X(\omega)H(\omega) \tag{2.11}$$

where $X(\omega)$, $H(\omega)$, and $Y(\omega)$ are the Fourier spectrum of $x(t)$, $h(t)$, and $y(t)$, respectively. The function $H(\omega)$ is defined as the *frequency response* of the system, and is one of the most fundamental properties of any system. For example, a *low-pass filter* has the frequency response as described in Figure 2.7.

FIGURE 2.6
LTI system.

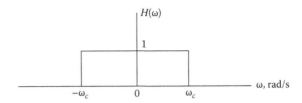

FIGURE 2.7
Frequency response of low-pass filter.

Example

Prove that $x(t) * \delta(t) = x(t)$.

Solution

According to the convolution theorem:

If $y(t) = x(t) * h(t)$

then,

$Y(\omega) = X(\omega)H(\omega)$

Since $h(t) = \delta(t)$, $H(\omega)=1$;

$\Rightarrow Y(\omega) = X(\omega)$

or, $y(t) = x(t)$

Note: *An extension of this property is* $x(t) * \delta(t-a) = x(t-a)$

2.4.2 Nonlinear Systems

Nonlinear systems are also widely used, most importantly in carrier systems, to transport information through communication channels. Generally, such a system performs a nonlinear operation such as *multiplication*, as compared with a linear operation like *addition* or *convolution*, as described in the previous section. Figure 2.8 shows such a nonlinear multiplier system. The input–output relation for a multiplier system is given as follows:

$$y(t) = x(t)h(t) \qquad (2.12)$$

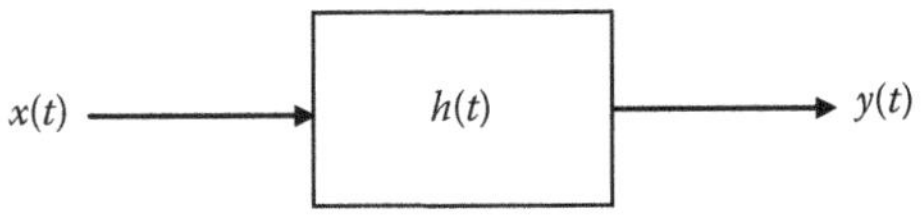

$x(t) \longrightarrow \boxed{\quad h(t) \quad} \longrightarrow y(t)$

FIGURE 2.8
Nonlinear multiplier system.

In this case, the relation between the input and output is a convolution (another example of duality principle!):

$$Y(\omega) = X(\omega) * H(\omega) \tag{2.13}$$

where $X(\omega)$, $H(\omega)$, and $Y(\omega)$ are the Fourier spectrum of $x(t)$, $h(t)$, and $y(t)$, respectively.

Example

Prove that $X(\omega) * \delta(\omega) = X(\omega)$.

Solution

According to the convolution theorem:

If $Y(\omega) = X(\omega) * H(\omega)$

then, $y(t) = x(t)\, h(t)$

Since $H(\omega) = \delta(\omega)$, $h(t)=1$;

or, $y(t) = x(t)$

Note: *An extension of this property is* $X(\omega) * \delta\,(\omega-a) = X(\omega-a)$

Example

A classic example of a multiplier system is in *carrier modulation* systems, which will be discussed in the next chapter. In such a system, the message signal $m(t)$ is multiplied with the carrier $x(t) = A \cos(\omega_c t)$. If $m(t)$ has a bandlimited spectrum, as shown in Figure 2.9, sketch the spectrum of the multiplied signal $y(t) = m(t)\, x(t)$.

Solution

$y(t) = x(t)m(t)$

$\Rightarrow y(t) = A \cos(\omega_c t)\, m(t)$

Applying the Fourier transform property from the previous section

$$Y(\omega) = M(\omega) * X(\omega)$$

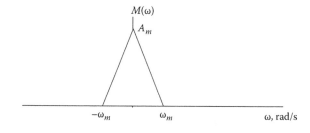

FIGURE 2.9
Bandlimited message spectrum.

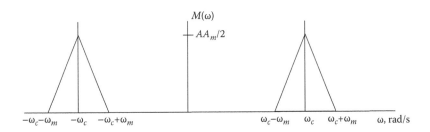

FIGURE 2.10
Spectrum of multiplied signal.

and using the Fourier transform property of a cosine function;

$$Y(\omega) = M(\omega) * \frac{A}{2}\left[\delta(\omega - \omega_c) + \delta(\omega + \omega_c)\right]$$

$$= \frac{A}{2}\left[M(\omega - \omega_c) + M(\omega + \omega_c)\right]$$

The spectrum of the multiplied signal is shown in Figure 2.10.

2.5 Practical Methods of Spectrum Analysis: DFT and IDFT

The discrete Fourier transform (DFT) is a practical extension of the Fourier transform, which is discrete both in time and frequency domains. Essentially, the infinite integrals, as defined in Equations 2.4 and 2.5, are converted into finite sums, with indices n and k used to represent time and frequency, respectively. The index n corresponds to the time value $t = n\Delta t$, s, where Δt is the sampling time interval. The index k corresponds to the frequency value $\omega = k\Delta\omega$, radians, where $\Delta\omega$ is the DFT output frequency interval. This property is used to divide the frequency interval $(0, 2\pi)$ into N points, to yield the DFT of the discrete-time sequence $x(n)$, $0 \le n \le N - 1$ as follows:

$$X(k) = X(\omega)\big|_{\omega = 2\pi k/N} = \sum_{n=0}^{N-1} x(n)e^{-j2\pi nk/N}, \quad 0 \le k \le N-1 \qquad (2.14)$$

The inverse discrete Fourier transform (IDFT) is given by the following equation:

$$x(n) = \frac{1}{N}\sum_{k=0}^{N-1} X(k)e^{j2\pi nk/N}, \quad 0 \le n \le N-1 \qquad (2.15)$$

A concise list of DFT transform properties is given in Table 2.3.

TABLE 2.3

DFT Theorems

Property	$f(n)$	$F(k)$
Periodicity in n and k	$x(n) = x(n \pm mN)$, for integer m	$X(k) = X(k \pm mN)$, for integer m
N-Point circular convolution	$x(n) \circledN h(n)$	$X(k) H(k)$
Circular time shift	$x((n-n_0)_N)$	$X(k) e^{-j2\pi n_0 k/N}$
Circular frequency shift	$e^{j2\pi n_0 k/N} x(n)$	$X((k-k_0)_N)$

Some of the key features and practical advantages of the DFT are as follows:

- The DFT maintains the time sequence $x(n)$ and the frequency sequence $X(k)$ as finite vectors having the same length N. Additionally, as seen from Equations 2.14 and 2.15, the DFT and IDFT are both finite sums, which make it very convenient to program these equations on computers and microprocessors.

- *Time–frequency relation:* This is a very important relation in practical DFT applications. The index n corresponds to the time value $t = n\Delta t$, s, where Δt is the sampling time interval. The index k corresponds to the frequency value $\omega = k\Delta\omega$, radians, where $\Delta\omega$ is the DFT output frequency interval. Then, for a given N-point DFT, the time frequency relation is given by

$$\Delta\omega = \frac{2\pi}{(N \Delta t)} \quad (2.16)$$

- The concept of *time shift* in the DFT is defined circularly: the sequence $x(n)$, $0 \le n \le N - 1$, is represented at N equally spaced points around a circle as shown in Figure 2.11a, for $N = 8$. Then a circular shift, represented as $x((n - 5)_8)$, for example, is implemented by moving the entire

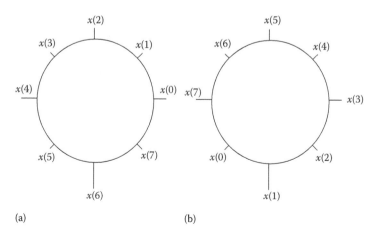

(a) (b)

FIGURE 2.11

Circular convolution process: (a) $x((n)_8)$ and (b) $x((n - 5)_8)$.

sequence $x(n)$ counterclockwise by 5 points, as illustrated in Figure 2.11b. Hence, the sequence $x(n) = [x(0)\ x(1)\ x(2)\ x(3)\ x(4)\ x(5)\ x(6)\ x(7)]$ becomes shifted sequence $x((n-5)_8) = [x(3)\ x(4)\ x(5)\ x(6)\ x(7)\ x(0)\ x(1)\ x(2)]$.

Example

Find the N point DFT of the following discrete-time sequence:

$$x(n) = a^n,\ 0 \le n \le N - 1$$

Solution

The N point DFT is given by

$$X(k) = \sum_{n=0}^{N-1} x(n)e^{-j2\pi nk/N}$$

$$= \sum_{n=0}^{N-1} a^n e^{-j2\pi nk/N}$$

$$= \sum_{n=0}^{N-1} (ae^{-j2\pi k/N})^n$$

which can be simplified using finite geometric series to give the final result:

$$X(k) = \frac{1-a^N}{1-ae^{-j2\pi k/N}},\quad 0 \le k \le N-1$$

2.6 Discrete-Time System Analysis: Circular Convolution

An N-point circular convolution of two sequences $x(n)$ and $h(n)$ is defined as

$$y(n) = x(n)\,\textcircled{N}\,h(n) = \sum_{m=0}^{N-1} x(m)h((n-m)_N),\quad 0 \le n \le N-1 \qquad (2.17)$$

Note that the sequences $x(n)$, $h(n)$, and $y(n)$ have the same vector length of N.

Example

Determine the circular convolution of the two 8-point discrete-time sequences, $x_1(n)$ and $x_2(n)$, given by

$$x_1(n) = x_2(n) = \begin{cases} 1, & 0 \le n \le 4 \\ 0, & 5 \le n \le 7 \end{cases}$$

Discuss the different methods of performing circular convolution.

Solution

The 8-point circular convolution is given by

$$x(n) = x_1(n) \, \textcircled{N} \, x_2(n)$$

$$= \sum_{m=0}^{7} x_1(m) x_2((n-m)_8)$$

Circular convolution can be carried out either by *analytic techniques*, such as the sliding tape method, or by *computer techniques* such as MATLAB®. We will discuss both approaches in the following:

Sliding tape method: This method can be done by hand calculation, if the number of points in the DFT, *N*, *is quite small*. The procedure is as follows:

- Write the sequences $x_1(m)$, $x_2(m)$, and $x_2((-m)_8)$ as shown in Table 2.4. The sequence $x_2((-m)_8)$ is obtained from the sequence $x_2(m)$, by writing the first element in the vector $x_2(m)$, then start with the last element in $x_2(m)$, and continue *backward*. Then the dot product of the vectors $x_1(m)$ and $x_2((-m)_8)$ gives the convolution output $x(0)$. Similarly, the next term in the table, $x_2((1-m)_8)$, is obtained by shifting $x_2((-m)_8)$ by one step to the right, and *back again to the beginning of the vector*. The dot product of the vectors $x_1(m)$ and $x_2((1-m)_8)$ gives the convolution output $x(1)$.
- Alternately, one could arrange the vector elements $x_1(m)$ and $x_2(m)$ in $N = 8$ equally spaced points around a circle, as shown in Figure 2.12a. The vector $x_2((-m)_8)$ is obtained by reflecting the vector elements of $x_2(m)$ about the horizontal axis as shown in Figure 2.12b. The vector $x_2((1-m)_8)$ is obtained by shifting the elements of the vector $x_2(m)$, by one position counterclockwise around the circle.

Hence, finally, the output vector is $x(n) = [2\,2\,3\,4\,5\,4\,3\,2]$.

Computer method: The circular convolution of the two sequences, $x_1(n)$ and $x_2(n)$, can also be obtained by using the convolution property of the

TABLE 2.4

Circular Convolution Procedure

$x_1(m) = [1\,1\,1\,1\,1\,0\,0\,0]$
$x_2(m) = [1\,1\,1\,1\,1\,0\,0\,0]$
$x_2((-m)_8) = [1\,0\,0\,0\,1\,1\,1\,1]; x(0) = 2$
$x_2((1-m)_8) = [1\,1\,0\,0\,0\,1\,1\,1]; x(1) = 2$
$x_2((2-m)_8) = [1\,1\,1\,0\,0\,0\,1\,1]; x(2) = 3$
$x_2((3-m)_8) = [1\,1\,1\,1\,0\,0\,0\,1]; x(3) = 4$
$x_2((4-m)_8) = [1\,1\,1\,1\,1\,0\,0\,0]; x(4) = 5$
$x_2((5-m)_8) = [0\,1\,1\,1\,1\,1\,0\,0]; x(5) = 4$
$x_2((6-m)_8) = [0\,0\,1\,1\,1\,1\,1\,0]; x(6) = 3$
$x_2((7-m)_8) = [0\,0\,0\,1\,1\,1\,1\,1]; x(7) = 2$

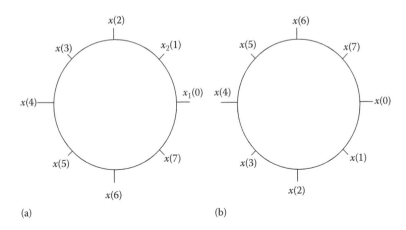

FIGURE 2.12
Circular convolution process: (a) $x_1((n)_8)$ and (b) $x_2((-n)_8)$.

DFT, which is listed as Property 2 in Table 2.3. This method consists of three steps:

Step 1: Obtain the 8-point DFTs of the sequences $x_1(n)$ and $x_2(n)$:

$$x_1(n) \rightarrow X_1(k)$$

$$x_2(n) \rightarrow X_2(k)$$

Step 2: Multiply the two sequences $X_1(k)$ and $X_2(k)$:

$$X(k) \rightarrow X_1(k)X_2(k), \quad \text{for } k = 0,1,2,\ldots,7.$$

Step 3: Obtain the 8-point IDFT of the sequence $X(k)$, to yield the final output $x(n)$:

$$X(k) \rightarrow x(n), \quad \text{for } n = 0,1,2,\ldots,7.$$

A brief MATLAB program to implement the aforementioned procedure is given in the following:

```
% Matlab program for circular convolution
clear;
x1 = [1 1 1 1 1 0 0 0];        sequence x₁(n)
x2 = [1 1 1 1 1 0 0 0];        sequence x₂(n)
X1 = fft(x1);                  DFT of x₁(n)
X2 = fft(x2);                  DFT of x₂(n)
X = X1.*X2;                    DFT of x(n)
x = ifft(X);                   IDFT of X(k)
```

Note: *MATLAB automatically utilizes a radix-2 FFT if N is a power of 2. If N is not a power of 2, then it reverts to a non-radix-2 process.*

2.7 Fast Fourier Transform

The fast Fourier transform, or the FFT as it is popularly termed, is probably the single most famous computer program in the field of Electrical Engineering, and represents the most practical version of the Fourier Transform, which is what we initially started out with. It is essentially a much faster computation method of the DFT, which was discussed in the previous section. The exceptional computational efficiency of the FFT is achieved by utilizing some periodic properties of the exponential functions in Equations 2.14 and 2.15. Some key properties of the FFT are given in the following text:

- An N-point DFT or N-point IDFT requires N^2 complex multiplications if computed directly from Equations 2.14 and 2.15, respectively. However, the same computation can be done with only $N \, Log_2 N$ complex multiplications, when a *radix*-2 (N is a power of 2) FFT is used. This is especially significant for large values of N: when $N = 128$, the number of complex multiplications is 16,384 for direct computation of DFT, and only 896 for a *radix*-2 FFT computation.

- FFT algorithms also exist when N is not a power of 2. These algorithms are called *non-radix*-2 FFT.

Example

A signal $f(t)$ is bandlimited at 1 KHz. The Fourier transform of the signal is obtained using an N point radix-2 FFT. What is the minimum number of points, N, required in order to obtain an output frequency resolution of 50 Hz?

Solution

The time–frequency relation is given in Equation 2.16 as

$$\Delta\omega = \frac{2\pi}{(N \, \Delta t)}$$

Sampling interval $\Delta t = 1/$Sampling frequency, f_s

$\qquad = 1/(2 \times$ signal bandwidth), to satisfy the

$\qquad$ sampling theorem

$\qquad = 1/(2 \times 1 \text{ kHz}) = 0.5 \text{ ms}.$

Output frequency resolution $= \Delta\omega = 2\pi \times \Delta f = 2\pi \times 50$

Substituting Δt and $\Delta\omega$ in the time–frequency relation, and solving for N:

$$N = 40$$

Since the problem specifies a radix-2 FFT, the closest power of 2 is $N = 64$.

2.8 Computation of Fast Fourier Transform with MATLAB®

The FFT was a major breakthrough in the efficient and fast computation of the Fourier transform of speech, music, and other fundamental signals. However, while the FFT is a very general formulation, there are some important points to keep in mind, when utilizing the FFT on *periodic* and *nonperiodic* signals.

- FFT evaluation of periodic signals

  Step 1: Sample the signal $x(t)$, as shown in Figure 2.13a, *over 1 period* of the signal, $T = 2\pi/\omega_0$, where ω_0 is the angular frequency of the signal. The sampling interval is

  $$\Delta t = \frac{T}{N}$$

  where N is the number of points in the FFT.

  Step 2: Generate the sampled signal $x(n)$, $n = 0,1,\ldots, N-1$. The input signal is stored as a vector $x = [x(0), x(1),\ldots, x(N-1)]$

  Step 3: The frequency interval is

  $$\Delta\omega = \frac{2\pi}{N\Delta t}$$

  $$= \omega_0$$

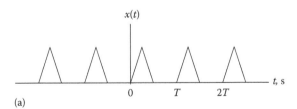

(a)

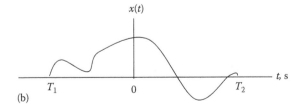

(b)

FIGURE 2.13
(a) Periodic signal and (b) nonperiodic signal.

Hence, the spectrum will appear at intervals of the fundamental frequency, which is true for periodic signal, as is shown by the Fourier series expansion. The program can be written as follows:

```
% MATLAB program to compute FFT of a periodic or
nonperiodic signal
X = fft(x);          calculates the FFT X(k) of the
                     vector x(n)
Xs = fftshift (X);   shifts the vector X(k) in
                     symmetric form
Xsm = abs(Xs);       magnitude spectrum
Xsp = angle(Xs);     phase spectrum
```

- FFT evaluation of nonperiodic signals

  Step 1: Sample the signal $x(t)$, shown in Figure 2.13b, *over the complete range* of the signal in the interval $T_1 \le t \le T_2$. The sampling interval is

  $$\Delta t = \frac{(T_2 - T_1)}{N}$$

  where N is the number of points in the FFT.

  Step 2: Generate the sampled signal $x(n)$, $n = 0,1, \ldots, N-1$. The input signal is stored as a vector $x = [x(0), x(1), \ldots, x(N - 1)]$

  Step 3: The frequency interval is

  $$\Delta \omega = \frac{2\pi}{(N \Delta t)}$$

  The MATLAB program for FFT computation is identical to the one given in the previous section, for periodic signals.

Problem Solving

P2.1 Determine the exponential Fourier series coefficients X_n of the following periodic signal:

$$x(t) = 5 + 3 \cos(t) + \sin(2t) + \sin(3t) - 0.5 \cos(5t + 60°)$$

(a) Approximately sketch the amplitude spectrum $|X_n|$ and phase spectrum/X_n.

(b) Find the power and rms value of the signal $x(t)$.

P2.2 The input $x(t)$ and the output $y(t)$ of a certain nonlinear channel are related as

$$y(t) = 5 \, x^2(t)$$

(a) Find the output signal $y(t)$ and its spectrum $Y(f)$ if the input signal is

$$x(t) = (1000/\pi) \text{ sinc } (1000t).$$

(b) Verify that the bandwidth of the output signal is twice that of the input signal.

P2.3 The frequency response of a certain class of digital filters called binomial filters is written as

$$H_r(\omega) = 2^N \left[\sin\left(\frac{\omega}{2}\right) \right]^r [\cos(\omega/2)]^{N-r}$$

in the range $-\pi \le \omega \le \pi$. Selecting $N = 2$, approximately sketch the magnitude response of the filters in the range $0 \le \omega \le \pi$ for the following cases:

(a) $r = 0$

(b) $r = 1$

(c) $r = 2$

P2.4 Determine the DFT, $X(k)$, of each of the following discrete-time sequences $x(n)$, for length $N = 8$:

(a) $x(n) = \delta(n)$

(b) $x(n) = 1, n$ even

        $= 0, n$ odd

(c) $x(n) = 0.5^n$

Approximately sketch the magnitude and phase of $X(k)$, for the range $k = 0,7$.

P2.5 (a) Let $x(n) = [1 -2 5 3 2]$, a sequence of length 5. Write down the following shifted sequences:

(i) $x[(n-1)_5]$

(ii) $x[(n+1)_5]$

(iii) $x[(-n)_5]$

(b) Let $x(n)$ and $h(n)$ be two finite-length sequences as given in the following:

$$x(n) = [-1\ 2\ -5], h(n) = [2\ -1\ -2]$$

Compute the circular convolution:

$$y(n) = x(n) ③ h(n)$$

P2.6 Suppose we have two 4-point sequences $x(n)$ and $h(n)$ as follows:

$$x(n) = \cos\left(\frac{\pi n}{2}\right), n = 0, 1, 2, 3$$

$$h(n) = 2^n, n = 0, 1, 2, 3$$

(a) Calculate the four-point DFT $X(k)$.

(b) Calculate the four-point DFT $H(k)$.

(c) Calculate $y(n) = x(n) \,\textcircled{4}\, h(n)$ by doing the circular convolution directly.

(d) Calculate $y(n)$ of part (iii) by multiplying the DFTs of $x(n)$ and $h(n)$ and performing an inverse DFT.

P2.7 The output of an LTI discrete-time system is given by

$$y(n) = x(n) * h(n)$$

where
$x(n)$ is the input
$h(n)$ is the impulse response of the system
* denotes circular convolution

(a) Using the convolution property of the DFT, write down a procedure for obtaining $y(n)$, given $x(n)$ and $h(n)$.

(b) If the convolution was performed using N-point DFTs and IDFTs, determine the number of complex multiplications required.

(c) If the convolution was performed using *radix*-2 FFTs and IFFTs, determine the number of complex multiplications required.

(d) Compare the results of parts (b) and (c) for $N = 32$.

Computer Laboratory

C2.1 Simulation of harmonic distortion in signal generators—Use of the FFT.

In this laboratory, the frequency spectrum of periodic signals at the output of signal generator is studied analytically and by experiment. The periodic signals shown in Figure 2.14 are considered. There are several useful commands in MATLAB, to generate periodic signals, and some examples are given.

```
Periodic square pulse
>> y =                    generates a square wave vector y with
   A*square(2*pi*f*t);    peak amplitude A and frequency f Hz.
                          The elements of y are calculated at the
                          time instances of the vector t.
>> y = A*square          generates a square wave vector, with
   (2*pi*f*t,duty);       identical parameters as above, but with
                          specified duty cycle. The duty cycle,
                          duty, is the percentage of the period
                          in which the signal is positive.
```

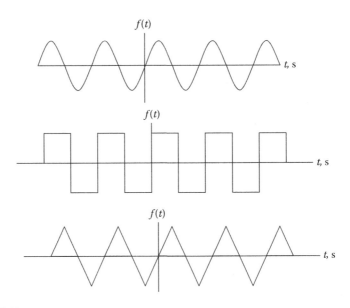

FIGURE 2.14
Periodic waveforms.

```
Periodic triangular pulse
>> y = A* tripuls(t);     generates samples of a continuous
                          triangle wave at the points specified
                          in array t, centered at t = 0. By
                          default, the triangle is symmetric and
                          has a duration of 1 sec.
>> y = tripuls(t,w);      generates a triangle, with parameters
                          as above, but duration of w, sec.
```

Note: Please also try other signal generation commands such as sin, cos, chirp, diric, gauspuls, pulstran, and rectpuls.

Set the frequency of the three signals to 1 MHz and peak amplitude to 1 V. Compute the FFT of each of the periodic signals using an output resolution of $\Delta f = 1$ MHz. Obtain the exponential Fourier series coefficients c_n, $1 \le n \le 5$ (from the FFT) for each of the waveforms mentioned. The power contained in the Fourier coefficients is given as follows:

$$Pc_n, comp = |c_n|^2 \text{ (mW)}$$

Hardware Laboratory

H2.1 Measurement of harmonic distortion in signal generators

Connect the output of the HP 3324A synthesized generator to the input of the HP 8590 L signal analyzer. Set the frequency of the generator to 1 MHz and peak amplitude to 1 V. Measure the power spectrum (dBm) for each of the signals to include the fundamental and first four harmonics.

(a) Compare the measured and simulated power spectrum (from the computer laboratory results) of the fundamental and first four harmonics in mW, after normalizing the peak values of the fundamental to 1 mW (0 dBm).

(b) Compute the % error between the computed and measured power spectrum (mW). The % error is defined as

$$\% \, error = \frac{|Pc_n, comp - Pc_n, meas|}{Pc_n, comp}$$

3

First-Generation Systems:
Analog Modulation

In the previous chapter, we introduced the concept of a *strong carrier signal* that is required to transport data and information through a medium such as coaxial, fiber-optic cable, satellite, or wireless. The process of adding or "piggybacking" the information to the carrier is termed as *modulation*. The carrier is a high-power, high-frequency signal, as compared to the information, which is low-power and low-frequency signal.

3.1 Amplitude Modulation

Amplitude modulation (AM) is perhaps the simplest type of modulation system, and some aspects of it are also used in systems where the main principle is one of the other kinds of modulation, perhaps even a digital type. It is the easiest to understand, because any *analog multiplication* of two signals produces a spectrum that contains the sum and the difference of the two frequencies. The following example illustrates this simple multiplication process.

Example

(a) Find the signal frequencies contained in the product, $y(t) = x(t)$ $m(t)$ of the following two signals:

Carrier signal: $x(t) = 50 \cos(1,000,000t)$

Information signal: $m(t) = 2 \cos(10,000t)$

(b) Determine the power and bandwidth of $y(t)$

Solution

(a) $y(t) = x(t)\, m(t) = 50 \cos(1,000,000t) \times 2 \cos(10,000t)$

Using trigonometric identity:

$y(t) = 50[\cos(1,000,000t + 10,000t) + \cos(1,000,000t - 10,000t)]$

$= 50[\cos(1,010,000t) + \cos(990,000t)]$

Hence, the two frequencies contained in y(t) are 1,010,000 rad./s and 990,000 rad./s.

(b) Power of $y(t) = 50^2/2 + 50^2/2 = 2500$ W.

Bandwidth of $y(t) = $ Highest frequency $-$ Lowest frequency

$$= 1,010,000 - 990,000$$

$$= 20,000 \text{ rad/s or } 3183.1 \text{ Hz}$$

3.1.1 Double Sideband Modulation

The multiplication process between the two signals (carrier and information) is shown in   Figure 3.1, which represents an AM transmitter. The carrier signal, $x(t)$, and its spectrum, $X(\omega)$, are shown in Figure 3.2a and b, respectively.

Likewise, a random message signal, $m(t)$, and its spectrum, $M(\omega)$, are shown in Figure 3.3a and b respectively. For any real signal, $m(t)$, it can be shown that the spectrum $M(\omega)$ is *symmetric*; hence the positive and negative halves of the spectrum are mirror images of each other. This property is very important in reducing bandwidth in modulation systems, since commonly used signals, such as audio (speech, music) and video, are real signals.

The multiplied output, or AM signal, is given by

$$y_{AM}(t) = A \, m(t) \cos(\omega_c t) \tag{3.1}$$

which is shown in Figure 3.4a, corresponding to the message signal in Figure 3.3a.

The amplitude of the carrier varies according to the amplitude of the message signal in Figure 3.3a, hence the term *AM*.

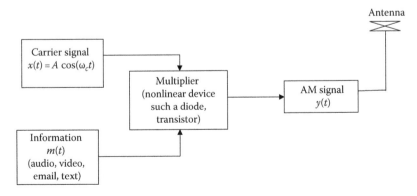

FIGURE 3.1
Block diagram of an AM transmitter.

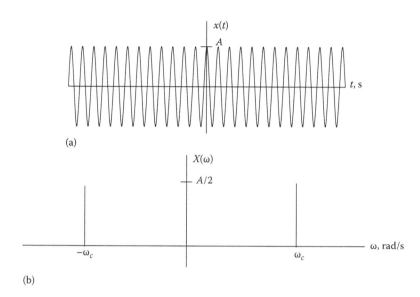

FIGURE 3.2
(a) Carrier signal and (b) spectrum of carrier signal.

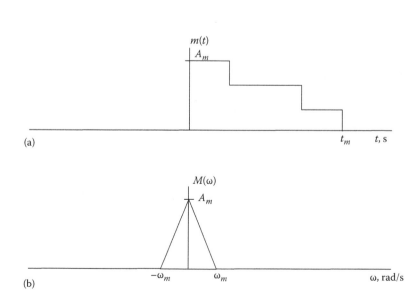

FIGURE 3.3
(a) Message signal and (b) spectrum of message signal.

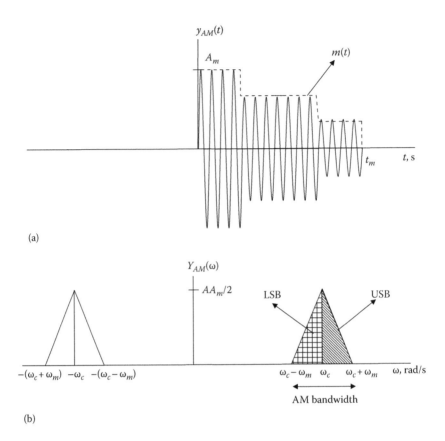

(a)

(b)

FIGURE 3.4
(a) AM signal and (b) spectrum of AM signal.

Referring to the modulation property from Chapter 2, the spectrum of $y(t)$ can be written as follows:

$$Y_{AM}(\omega) = M(\omega) * \frac{A}{2}\left[\delta(\omega - \omega_c) + \delta(\omega + \omega_c)\right]$$

$$= \frac{A}{2}\left[M(\omega - \omega_c) + M(\omega + \omega_c)\right] \tag{3.2}$$

The spectrum of the multiplied signal is shown in Figure 3.4b. The two halves of the spectrum are named as USB (upper sideband) and LSB (lower sideband), each occupying half the total bandwidth (ω_m rad/s), and they are shown in different shading patterns. The property of the negative side of the spectrum is identical to the positive side, and this process is termed as *Conventional AM or Double Sideband AM.*

Thus, in the *AM transmitter*, through a process of signal multiplication, the message signal is upconverted to a higher frequency, while maintaining the

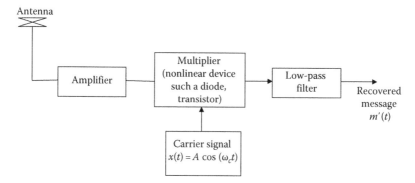

FIGURE 3.5
Block diagram of an AM receiver.

message signal spectrum. Correspondingly, in the *AM receiver*, the upconverted signal has to be processed to recover the message signal. This process is shown in Figure 3.5. The amplifier is essential to boost the received AM signal, which is usually very weak after traveling a long distance through the *channel*. Interestingly, to recover the message signal, we multiply the received AM signal with the carrier signal! This multiplication yields:

$$z(t) = \text{Received AM signal} \times \text{carrier signal}$$

$$= Am(t)\cos(\omega_c t) \times \cos(\omega_c t)$$

$$= Am(t)\cos^2(\omega_c t)$$

$$= Am(t)\left[\frac{1+\cos(2\omega_c t)}{2}\right]$$

$$= Am(t)\left[\frac{1}{2}\right] + Am(t)\left[\frac{\cos(2\omega_c t)}{2}\right] \tag{3.3}$$

The first term in Equation 3.3 is a *low-frequency* message signal, while the second term is the *high-frequency* component, which can be removed by the subsequent *low-pass filter*. The modulation process is now complete. Some important AM parameters are described as follows:

- *Power* in the AM signal defined by Equation 3.3 is

$$P_{AM} = \lim_{T=\infty} \frac{1}{T} \int_0^T |y_{AM}(t)|^2 dt$$

$$= \lim_{T=\infty} \frac{1}{T} \int_0^T |A\, m(t)\cos(\omega_c t)|^2 dt \tag{3.4}$$

For a real signal $m(t)$, Equation 3.4 reduces to the following form:

$$P_{AM} = \lim_{T=\infty} \frac{1}{T} \int_0^T \left[Am(t)\cos(\omega_c t) \right]^2 dt$$

$$= \lim_{T=\infty} \frac{1}{T} \int_0^T \left[Am(t)\cos(\omega_c t) \right]^2 dt$$

$$= \lim_{T=\infty} \frac{1}{T} \int_0^T A^2 m^2(t) \left[\frac{1+\cos(2\omega_c t)}{2} \right] dt$$

$$= \lim_{T=\infty} \frac{A^2}{2T} \int_0^T m^2(t)\, dt + \frac{A^2}{2T} \int_0^T m^2(t)\cos(2\omega_c t)\, dt$$

$$= \lim_{T=\infty} \frac{A^2}{2T} \int_0^T m^2(t)\, dt \tag{3.5}$$

and the second term averages to zero, as it is a high-frequency term. The power equation represents the *sideband power* P_s. In some modulation schemes, if the carrier is also sent as a reference, then the *carrier power*, P_c, must also be added, and the total power becomes

$$P_{AM} = P_c + P_s$$

$$= \frac{A^2}{2} + \lim_{T=\infty} \frac{A^2}{2T} \int_0^T m^2(t)\, dt \tag{3.6}$$

- *Bandwidth* of the AM signal, as seen from Figure 3.4b, is $(\omega_c + \omega_m) - (\omega_c - \omega_m) = 2\omega_m$.
- *Modulation index*, μ, is an important design parameter in AM systems, and defined as

$$\mu = \frac{m(t)\big|_{max}}{A}$$

where $-1 < \mu < 1$, for avoiding distortion in the AM signal.

- *Efficiency*, η, of the AM system is given by

$$\eta = \frac{\text{Sideband power}}{\text{Carrier power} + \text{Sideband power}} = \frac{P_s}{P_c + P_s}$$

Example

For the message signal shown in Figure 3.6, with $\mu = 0.8$, find the following parameters, assuming *conventional AM modulation*:

(a) The amplitude and power of the carrier
(b) The sideband power and power efficiency η

Solution

(a) Modulation index $\mu = \dfrac{m(t)|_{max}}{A} = \dfrac{5}{A} = 0.8$ (as given)

$\Rightarrow A = 6.25$ V

$$\text{Power in carrier } P_c = \frac{A^2}{2} = \frac{6.25^2}{2}$$

$$= 19.53 \text{ W}$$

(b) Sideband power

$$P_s = \lim_{T=\infty} \frac{A^2}{2T} \int_0^T m^2(t)\,dt$$

$$= \frac{6.25^2}{2(4\,\text{ms})} \int_0^{4\text{ms}} m^2(t)\,dt$$

$$= \frac{6.25^2}{2(4\,\text{ms})}\Big[5^2(1\text{ms}) + 3^2(2\text{ms}) + 1^2(1\text{ms})\Big]$$

$$= 214.84 \text{ W}$$

Efficiency

$$\eta = \frac{\text{Sideband power}}{\text{Carrier power} + \text{Sideband power}} = \frac{P_s}{P_c + P_s}$$

$$\frac{214.84}{19.53 + 214.84} = 0.9167 \text{ or } 91\%.$$

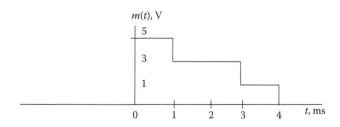

FIGURE 3.6
Message signal for an AM system.

3.1.2 Single Sideband Modulation

Variations of the conventional AM system, discussed in the previous section, include Single Sideband Modulation (SSB). In SSB, only *one of the two side-bands* of the AM signal is transmitted, to save bandwidth. This is permissible since, as pointed out in the earlier section, both sidebands of a real message signal are symmetrical; hence, *one can be recovered from the other.* Figure 3.7a and b shows the spectrum of upper and lower SSB systems, respectively. Some important SSB parameters are described as follows:

- *Power* of the SSB signal is *half* that of conventional AM signals, since only one sideband is transmitted.
- *Bandwidth* of the SSB signal, as seen from Figure 3.7a and b, is ω_m; hence, there is also a 50% saving in bandwidth as compared to conventional AM signal.

Example

For the message signal shown earlier in Figure 3.6, with $\mu = 0.8$, find the following parameters, assuming *SSB modulation*:

(a) The amplitude and power of the carrier
(b) The sideband power and power efficiency η

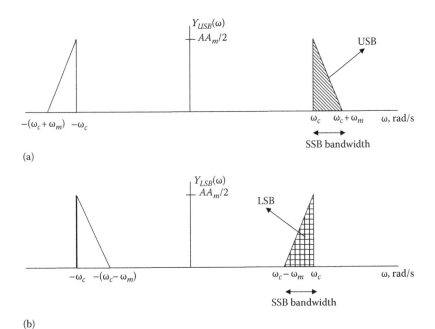

(a)

(b)

FIGURE 3.7
(a) Spectrum of USB signal and (b) spectrum of LSB signal.

Solution

(a) Modulation index $\mu = \dfrac{m(t)|_{max}}{A} = \dfrac{5}{A}$

$$= 0.8 \,(\text{as given})$$

$\Rightarrow A = 6.25$ V

$$\text{Power in carrier } P_c = \frac{A^2}{2} = \frac{6.25^2}{2}$$

$$= 19.53 \text{ W}$$

(b) Sideband power

$$P_s = \lim_{T=\infty} \frac{A^2}{4T} \int_0^T m^2(t)\,dt, \text{ since there is only one sideband;}$$

$$= \frac{6.25^2}{4(4\,\text{ms})} \int_0^{4\text{ms}} m^2(t)\,dt$$

$$= \frac{6.25^2}{4(4\,\text{ms})} \left[5^2(1\,\text{ms}) + 3^2(2\,\text{ms}) + 1^2(1\,\text{ms}) \right]$$

$$= 107.42 \text{ W}$$

Efficiency

$$\eta = \frac{\text{Sideband power}}{\text{Carrier power} + \text{sideband power}} = \frac{P_s}{P_c + P_s}$$

$$\frac{107.42}{19.53 + 107.42} = 0.8462 \text{ or } 84.62\%.$$

Hence, the efficiency actually decreases; however, power and bandwidth have been saved.

SSB signals can be generated by using a *band-pass filter*, which allows the USB or LSB signal to pass through, as will be explained now. Consider the spectrum, $M(\omega)$, of a band-limited message real signal $m(t)$, as shown in Figure 3.8. Assuming $m(t)$ to be real, the spectrum is hence symmetrical; let us call the *positive* and *negative* parts of the spectrum as $M_+(\omega)$ and $M_-(\omega)$, respectively. This is also illustrated in Figure 3.8. The message signal spectrum can then be written as follows:

$$M(\omega) = M_+(\omega) + M_-(\omega) \tag{3.7}$$

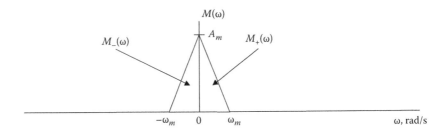

FIGURE 3.8
Positive and negative parts of message spectrum.

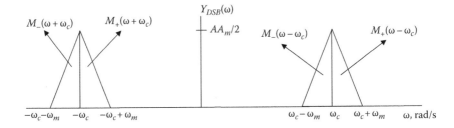

FIGURE 3.9
Components of modulated spectrum.

Then, multiplying the message signal with the carrier $A \cos(\omega_c t)$ yields the modulated double-sideband spectrum $Y_{DSB}(\omega)$, as shown in Figure 3.9, with the shifted message spectrum components. When this modulated signal is passed through the band-pass filter $H_{hp}(\omega)$, as shown in Figure 3.10a, it yields the *USB signal* shown in Figure 3.7a. Likewise, when the modulated signal is passed through the band-pass filter $H_{hp}(\omega)$ in Figure 3.10b, it yields the *LSB signal* shown in Figure 3.7b.

A more elegant way of SSB generation is a *Balanced Modulator*, which uses the principle of phase cancellation. To understand this, let us go back to Equation 3.4:

$$M(\omega) = M_+(\omega) + M_-(\omega)$$

Referring to Figure 3.7 and the definition of the SSB signal, the USB signal spectrum can be written as

$$Y_{USB}(\omega) = M_+(\omega - \omega_c) + M_-(\omega + \omega_c) \tag{3.8}$$

Adding and subtracting a few identical terms:

$$Y_{USB}(\omega) = M_+(\omega - \omega_c) + M_-(\omega + \omega_c) + 0.5\left[M_-(\omega - \omega_c) - M_-(\omega - \omega_c) \right]$$

$$+ 0.5\left[M_+(\omega + \omega_c) - M_+(\omega + \omega_c) \right]$$

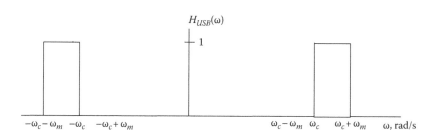

(a)

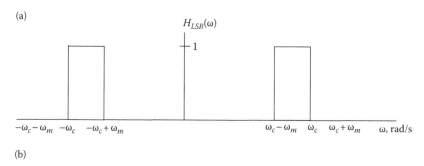

(b)

FIGURE 3.10
(a) USB band-pass filter and (b) LSB band-pass filter.

and rearranging:

$$Y_{USB}(\omega) = 0.5\left[M_+(\omega - \omega_c) + M_-(\omega - \omega_c)\right] + 0.5\left[M_+(\omega + \omega_c) + M_-(\omega + \omega_c)\right]$$

$$-0.5M_-(\omega - \omega_c) - 0.5M_+(\omega + \omega_c) + 0.5\left[M_+(\omega - \omega_c) + M_-(\omega + \omega_c)\right]$$

$$= Y_{USB1}(\omega) + Y_{USB2}(\omega) \tag{3.9}$$

$$Y_{USB1}(\omega) = 0.5\left[M_+(\omega) * \delta(\omega - \omega_c) + M_-(\omega) * \delta(\omega - \omega_c) + M_+(\omega) * \delta(\omega + \omega_c)\right.$$

$$\left. + M_-(\omega) * \delta(\omega + \omega_c)\right]$$

$$= 0.5\left[M_+(\omega) + M_-(\omega)\right] * \delta(\omega - \omega_c) + 0.5\left[M_+(\omega) + M_-(\omega)\right] * \delta(\omega + \omega_c)$$

$$= 0.5M(\omega) * \left[\delta(\omega - \omega_c) + \delta(\omega + \omega_c)\right] \tag{3.10}$$

The term $Y_{USB1}(\omega)$ can be generated by the multiplication factor $0.5m(t)\cos(\omega_c t)$; however, how do we generate the term $Y_{USB2}(\omega)$?

$$Y_{USB2}(\omega) = -0.5M_-(\omega - \omega_c) - 0.5M_+(\omega + \omega_c) + 0.5M_+(\omega - \omega_c) + 0.5M_-(\omega + \omega_c)$$

$$= 0.5\left[M_+(\omega - \omega_c) - M_-(\omega - \omega_c)\right] - 0.5\left[M_+ s(\omega + \omega_c) - M_-(\omega + \omega_c)\right]$$

$$= 0.5\left[M_+(\omega) - M_-(\omega)\right] * \delta(\omega - \omega_c) - 0.5\left[M_+(\omega) - M_-(\omega)\right] * \delta(\omega + \omega_c)$$

$$= 0.5M(\omega)\text{sgn}(\omega) * \left[\delta(\omega - \omega_c) - \delta(\omega + \omega_c)\right] \tag{3.11}$$

where sgn(ω) is the signum function defined as follows:

$$\text{sgn}(\omega) = \begin{cases} -1, & \omega > 0 \\ +1, & \omega < 0 \end{cases}$$

Hence, combining Equations 3.8 and 3.9, we obtain

$$Y_{USB}(\omega) = Y_{USB1}(\omega) + Y_{USB2}(\omega)$$

$$= 0.5M(\omega)*\left[\delta(\omega-\omega_c)+\delta(\omega+\omega_c)\right] + 0.5M(\omega)\text{sgn}(\omega)*\left[\delta(\omega-\omega_c)-\delta(\omega+\omega_c)\right]$$

(3.12)

In the time domain, this equation can be transformed as follows:

$$y_{USB}(t) = 0.5m(t)\cos(\omega_c t) + 0.5\hat{m}(t)\sin(\omega_c t) \tag{3.13}$$

where $\hat{m}(t)$ is defined as the Hilbert transform of the signal $m(t)$. Comparing Equations 3.12 and 3.13, we can define the following Fourier transform pair:

$$\hat{m}(t) \leftrightarrow jM(\omega)\text{sgn}(\omega) \tag{3.14}$$

The first term in Equation 3.13 can be generated by multiplying the signal $m(t)$ with the carrier, as in conventional DSB modulation, while the second

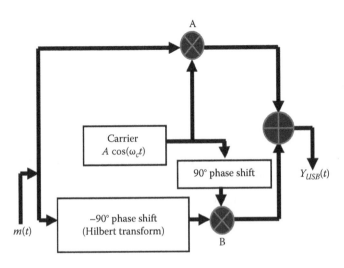

FIGURE 3.11
Single sideband (SSB) modulator.

term can be generated by a two-stage phase-shifting: shifting the phase of $m(t)$ by 90° to yield $\hat{m}(t)$, and then multiplying it with a carrier, shifted in phase by 90° too!

Basically, we need *two subsystems* to implement Equation 3.13: first subsystem to implement the product: $m(t)\cos(\omega_c t)$ and the second subsystem to implement the product: $\hat{m}(t)\sin(\omega_c t)$. The block diagram to implement this equation is shown in Figure 3.11. The first product is realized by the *multiplier A*, which is quite simple. The realization of the second product $\hat{m}(t)\sin(\omega_c t)$ by *multiplier B* is more complex: The 90° phase shifter basically converts the $\cos(\omega_c t)$ into $\sin(\omega_c t)$, and the −90° phase shifter implements the Hilbert transform given by Equation 3.14.

Example

A music signal with a bandwidth of 20 kHz is transmitted from an AM radio station, with a carrier power of 50 W and transmission frequency of 650 kHz. If the average music signal power is 3 W, determine the following:

(a) Power and bandwidth of the modulated signal, if conventional AM modulation is used
(b) Power and bandwidth of the modulated signal, if SSB modulation is used

Solution

(a) Carrier power is given as follows:

$$P_c = \frac{A^2}{2} = 50 \text{ W (given)}$$

$$\Rightarrow A = 10 \text{ V}$$

Hence, total power in conventional AM, from Equation 3.6, is

$$P_{AM} = P_c + P_s$$

$$= \frac{A^2}{2} + \lim_{T \to \infty} \frac{A^2}{2T} \int_0^T m^2(t)\,dt$$

$$= \frac{A^2}{2} + \frac{A^2}{2}\text{(signal power)}$$

$$= 50 \text{ W} + 50 \text{ W}(3 \text{ W})$$

$$= 200 \text{ W}$$

Bandwidth of conventional AM signal $= 2f_m = 2 \times 20 \text{ kHz} = 40 \text{ kHz}$

(b) Total power in SSB AM, from Equation 3.6, is

$$P_{AM} = P_c + P_s$$

$$= \frac{A^2}{2} + \lim_{T=\infty} \frac{A^2}{4T} \int_0^T m^2(t)\,dt$$

$$= \frac{A^2}{2} + \frac{A^2}{4}(\text{signal power})$$

$$= 50\ W + 50\ W(3\ W)/2$$

$$= 125\ W$$

Bandwidth of SSB signal $= f_m = 20$ kHz.

3.2 Angle Modulation

Angle modulation is more complex than AM, since it involves *changing the carrier frequency or phase*, rather than the carrier amplitude. It was proposed first by *Edwin Armstrong*, who had to fight an uphill battle, with intense opposition from the AM industry. Even then, it was accepted only after his lifetime. There are two forms of angle modulation: *Frequency Modulation* (*FM*) and *phase modulation* (*PM*).

3.2.1 Frequency Modulation

Basically in FM, the *frequency* of the carrier signal, $A \cos(\omega t)$, varies according to the message signal; the instantaneous frequency is given by

$$\omega = \omega_c + k_f\, m(t) \tag{3.15}$$

where k_f is the FM constant, unit (Hz/V). Since frequency and phase of a signal are related by the equation

$$\omega = \frac{d\theta}{dt} \tag{3.16}$$

it follows from Equations 3.15 and 3.16 that

$$\theta = \int_{-\infty}^{t} \omega \, dt$$

$$= \int_{-\infty}^{t} [\omega_c + k_f m(t)] \, dt$$

$$= \omega_c t + k_f \int_{-\infty}^{t} m(t) \, dt \qquad (3.17)$$

Finally, the FM signal can be defined by the following equation:

$$y_{FM}(t) = A \cos[\theta]$$

$$= A \cos \left[\omega_c t + k_f \int_{-\infty}^{t} m(t) \, dt \right] \qquad (3.18)$$

To illustrate the application of FM, let us consider the same message signal that was considered in the earlier AM section, as shown in Figure 3.12a. Then the FM signal, corresponding to this message signal, is shown in Figure 3.12b. As seen from Figure 3.12b, the frequency of the carrier signal is higher when the message signal amplitude is large, and is lower when the message signal amplitude is smaller. The variation in frequency of the FM signal is called the *maximum frequency deviation*, $\Delta\omega$. Comparing with Equation 3.18:

$$\Delta\omega = k_f \, |m(t)|_{max}$$

$$\text{or, } \Delta f = \frac{k_f |m(t)|_{max}}{2\pi} \qquad (3.19)$$

As in AM, the *modulation index* is an important parameter in FM too, and is defined as follows:

$$\beta = \frac{\Delta f}{B} \qquad (3.20)$$

where B is the bandwidth of the baseband message signal.

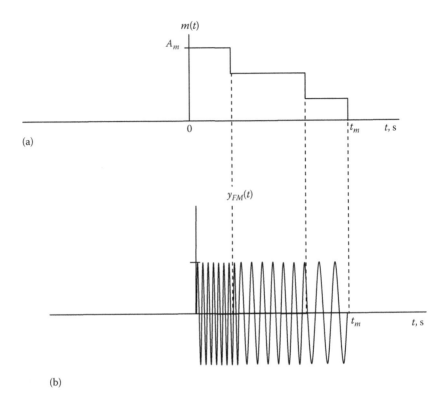

FIGURE 3.12
(a) Message signal and (b) FM signal.

3.2.2 Phase Modulation

In PM, the *phase* of the carrier signal, $A \cos(\omega_c t + \theta)$, varies according to the message signal; the instantaneous phase is given by

$$\theta = k_p m(t) \tag{3.21}$$

where k_p is the PM constant, unit (rad/V). Hence, the PM signal can be defined by the following equation:

$$y_{PM}(t) = A \cos\left[\omega_c t + \theta \right]$$
$$= A \cos\left[\omega_c t + k_p m(t) \right] \tag{3.22}$$

The PM signal looks exactly like the FM signal shown in Figure 3.12b, corresponding to the same baseband signal, since a shift in phase and a shift in frequency produce the same effect. The variation in phase of the PM signal is called the *maximum phase deviation,* $\Delta\theta$. Comparing with Equation 3.22,

$$\Delta\theta = k_p \, |m(t)|_{max}$$

Since, from Equation 3.16: $\omega = d\theta/dt$, the *maximum frequency deviation* in the PM signal is given by

$$\Delta\omega = k_p \left.\frac{dm(t)}{dt}\right|_{max} \quad \text{or} \quad \Delta f = \frac{k_p}{2\pi}\left.\frac{dm(t)}{dt}\right|_{max} \tag{3.23}$$

The *modulation index* of PM is defined exactly as in FM, given by Equation 3.20:

$$\beta = \frac{\Delta f}{B}$$

However, we should note that the maximum frequency deviation has a different formulation, as given in Equation 3.23. Some important FM parameters are described below:

- *Power* in the FM/PM signal is relatively easier to calculate; since it is a cosine wave with constant amplitude A, the average power is given by

$$P_{PM/FM} = \frac{A^2}{2} \tag{3.24}$$

- *Bandwidth* of the FM/PM signal is given by

$$B_{FM/PM} = 2(\Delta f + B)$$

or alternatively, in terms of modulation index as

$$B_{FM/PM} = 2B(1 + \beta) \tag{3.25}$$

where B is the bandwidth of the baseband message signal. However, keep in mind that the maximum frequency deviation, Δf, has different formulae from FM and PM systems, as given by equations.

The bandwidth of FM is inherently larger compared to the bandwidth of AM, hence accounting for its higher fidelity and quality of audio and music. However, the lower bandwidth of AM is useful in video, which occupies the MHz range as compared to audio, which occupies the lower kHz range. A classic application of AM and FM is in *television*, which uses FM for audio and AM for video.

Example

An FM system signal is given by

$s(i) = 2 \cos [10^7\pi t + 2 \sin (3500\pi t + 0.3\pi)]$

Find the power and bandwidth of this FM signal.

Solution

Power in FM signal = $A^2/2 = 2^2/2 = 2$ W.

Bandwidth of FM = $2(\Delta f + B)$

Δf = Maximum frequency deviation = $\max(d\theta/dt)$

Phase $\theta = 10^7\pi t + 2 \sin(3500\pi t + 0.3\pi)$

Hence, $d\theta/dt = 10^7\pi + 7000\pi \cos(3500\pi t + 0.3\pi)$

Hence, $\Delta f = 7000\pi/2\pi = 3500$ Hz.

Signal bandwidth B = Maximum signal frequency = $3500 \ \pi$ rad/s => 1750 Hz.

Hence, bandwidth of FM signal = $2(\Delta f + B) = 2(3500 + 1750) = 10{,}500$ Hz.

3.3 Comparison of AM and FM Modulation Systems

- FM is less susceptible to noise as compared to AM, since frequency of the carrier contains the message information and noise always affects the amplitude.

- FM has better sound quality than AM; however, this also requires higher bandwidth than AM. Hence, typically, AM is used for high-bandwidth video transmission, while FM is used for low-bandwidth audio transmission.

- Power requirement is less in FM as compared to AM, since the modulated signal has constant sinusoidal amplitude.

- Transmitter/receiver design is more complex in FM, as compared to AM.

- FM has shorter range of transmission, requiring line of sight (LOS) propagation, while AM signals can travel large distances, even across oceans.

3.4 Noise and Filtering in Analog Modulation Systems

In communication systems, noise is an *undesired* random disturbance that affects signals in the transmission channel. Noise can arise from *natural* causes such as *lightning, atmospheric, cosmic sources,* or from *man-made sources* such as *interference* from other signals using the same frequency band, and from *power lines* carrying high currents. In modern cellular channels, *fading* is also a form of signal distortion, which arises due to two primary reasons: *multipath* addition of signals at the receiver, and also *Doppler shift* in frequency due to mobile objects such as cars and other vehicles. Multipath occurs due to signals getting reflected, for example, from buildings and other obstacles and arriving at the receiver with phase differences that lead to rapid fluctuation of signal strength. Whatever be the cause of noise in the channel, the overall effect is quantified by the *signal-to-noise ratio (SNR)*, which is defined as

$$SNR = \frac{\text{Signal power(W)}}{\text{Noise power(W)}}$$

or in logarithmic units

$$SNR_{dB} = 10\log_{10}(SNR) \tag{3.26}$$

Additive white Gaussian noise (AWGN) is a basic noise model that is used in communication theory to mimic the effect of many random processes that occur in nature. The term "additive" refers to the nature of the noise, which basically adds itself to the modulated signal; the term "white" refers to the fact that this noise has uniform power across the entire frequency band of the communication system. The *Gaussian distribution* is a very popular model to explain many random processes, including noise in communication channels. It obeys the *normal probability distribution*, with an average time domain value of zero. For a *random signal y(t)*, such as the combination of an AM or FM signal with noise, the probability distribution function is given by

$$P(y) = \frac{1}{\sigma\sqrt{2\pi}} e^{-(x-\mu)^2/2\sigma^2} \tag{3.27}$$

where
  μ is the mean
  σ is the standard deviation of the random signal

A standard normal distribution is a special case of Equation 3.27 when $\mu = 0$ and $\sigma = 1$.

3.4.1 Noise Performance of AM and FM Circuits

A major advantage of FM in a communications circuit, as compared with AM, is the possibility of improved SNR; for typical voice communications channels, improvements are typically 5–15 dB. Basically, since FM signals have constant amplitude, *limiter circuits* can remove excess noise; however, limiters cannot be applied to AM signals, which have varying amplitude.

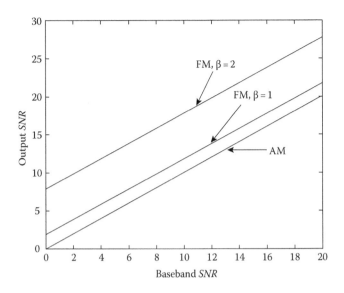

FIGURE 3.13
Output SNR performance for AM and FM systems.

Additional techniques such as *pre-emphasis* of higher audio frequencies with corresponding *de-emphasis* in the receiver, can be used to improve overall SNR in FM circuits. The output SNR is given in Equation 3.28, below for different AM and FM systems, in terms of the baseband or message signal SNR.

$$SNR_{AM} = SNR_{baseband}$$

$$SNR_{FM} = 1.5 \ \beta^2 SNR_{baseband} \tag{3.28}$$

The key point is that while the output *SNR* in AM is constant, the output FM *SNR* can be improved by increasing the modulation index β. For example, if the modulation index is doubled, then output *SNR* is increased by *four times*. This improvement in FM SNR, as compared to AM SNR, is shown in Figure 3.13, and also the increased SNR gain with higher values of β(beta).

3.4.2 Filtering Techniques to Minimize Noise Effects in Communication Channels

Different filtering mechanisms have been developed over the years for the primary purpose of reducing noise content in AM and FM signals. One of the most generalized approaches to noise filtering was developed by *Norbert Weiner*, which is based on the mean-squared error minimization between the transmitted signal and the received signal. The block diagram of a

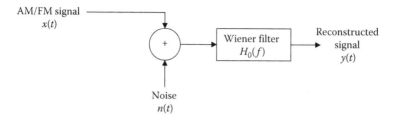

FIGURE 3.14
Communication system with Wiener filtering.

communication system, with filtering, is shown in Figure 3.14. The optimum Wiener filter has the following transfer function:

$$H_o(f) = \frac{S_x}{S_x + S_n} \tag{3.29}$$

where S_y and S_n are the *power spectral density* (*psd*) of the signal and the noise, respectively, with units of Watts/Hz. For *deterministic* signals, like AM and FM, the psd is defined by the equation:

$$S_x(f) = \underset{T \to \infty}{\mathrm{Lim}} \frac{1}{2T} \left| \int_{-T}^{T} x(t) e^{-j2\pi ft} dt \right|^2 \tag{3.30}$$

Then, the total power in the signal is given by

$$P_x = \int_{-\infty}^{\infty} S_x(f) df$$

For random signals like the channel noise, the psd is usually estimated by measurements. White noise, for example, has a constant psd of N_0 W/Hz.

- *Advantages* of Wiener filtering include the fact that it is a very generalized approach to noise filtering in communication channels. Wiener filtering can also be extended to *two-dimensional image processing*, to improve the quality of pictures; for example, deep-space images corrupted with atmospheric turbulence have been successfully recovered by Wiener filtering.

- However, Wiener filtering applies only to *stationary* processes, that is, channels which do not change significantly with time. However, there are channels that are *nonstationary*, such as cellular channels, which keep changing due to *mobile* activity. In these types of channels, filtering has to be *adaptive*, or adjusting to the changing channel profile. *Digital filters*, unlike analog filters, can be adaptive, as will be detailed in the next chapter.

The following example will illustrate the SNR improvement in a communication signal due to Wiener filtering.

Example

An AM communication signal has a psd of $S_x(f) = 2\Pi\big((|f| - 740\,\mathrm{K})/20\,\mathrm{K}\big)$,
where Π is the pulse function, defined earlier in Chapter 2. This signal
passes through a noisy channel with a constant psd of $S_n(f) = 10^{-3}\,\mathrm{W/Hz}$,
for $|f| \leq 740$ kHz, and zero elsewhere. The noisy signal is then passed
through a Wiener reconstruction filter, $H_o(f)$:

(a) Plot the psd of the input signal and noise, respectively.
(b) Calculate the input SNR, at the input of the filter.
(c) Calculate the SNR improvement at the output of the Wiener
filter.

Solution

(a) The psd plots of the signal and noise are given in Figure 3.15a
and b, respectively.

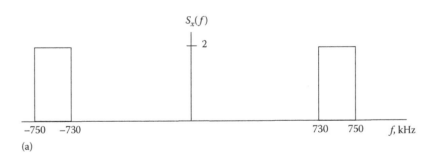

(a)

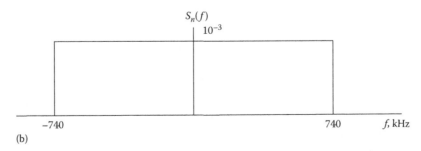

(b)

FIGURE 3.15
PSDs of signal and noise.

(b) The SNR at the input to the filter is given by

$$\text{SNR}_i = \frac{S_i}{N_i}$$

$$= \frac{\displaystyle\int_{-\infty}^{\infty} S_x(f)\,df}{\displaystyle\int_{-\infty}^{\infty} S_n(f)\,df} = \frac{2(2)(20\,\text{K})}{10^{-3}(2)(740\,\text{K})} = 54.05$$

(c) The Wiener reconstruction filter response is given by

$$H_o(f) = \frac{S_x}{S_x + S_n}$$

$$= \frac{2\Pi\left(\dfrac{|f| - 740\,\text{K}}{20\,\text{K}}\right)}{2\Pi\left(\dfrac{|f| - 740\,\text{K}}{20\,\text{K}}\right) + 10^{-3}\,\Pi\left(\dfrac{f}{1480K}\right)}$$

$$= \begin{cases} \dfrac{2}{2.001}, & 730\,\text{K} \le |f| \le 740\,\text{K} \\[2mm] 1, & 740\,\text{K} \le |f| \le 750\,\text{K} \end{cases}$$

and the SNR at the output of the filter, by examining Figure 3.15a and b, can be written as

$$\text{SNR}_o = \frac{S_o}{N_o}$$

$$= \frac{\displaystyle\int_{-\infty}^{\infty} S_x(f)\,|H_o(f)|^2\,df}{\displaystyle\int_{-\infty}^{\infty} S_n(f)\,|H_o(f)|^2\,df}$$

$$= \frac{2(2)\left(\dfrac{2}{2.001}\right)^2(10\,\text{K}) + 2(2)(1)^2(10\,\text{K})}{10^{-3}(2)\left(\dfrac{2}{2.001}\right)^2(10\,\text{K})} = 4002.0$$

Hence, output SNR is much higher than the input SNR, by a factor of 4002/54.05 = 74.04, or 18.6 dB.

Problem Solving

P3.1 Given the message signal $m(t) = 5 \sin(100\pi t)$ and the carrier wave-form is: $2 \cos(2000\pi t)$, approximately sketch the following:

(a) $s_{AM}(t)$ (approximate graph)
(b) The Fourier spectrum of $s_{AM}(t)$
(c) $s_{SSB}(t)$ (approximate graph)
(d) The Fourier spectrum of $s_{SSB}(t)$

P3.2 For the modulating signal $m(t) = 10 \cos(50 \pi t)$, and system efficiency $\mu = 0.8$, determine the following:

(a) The amplitude and power of the carrier
(b) The sideband power and power efficiency η for $s_{AM}(t)$
(c) The sideband power and power efficiency η for $s_{SSB}(t)$

P3.3 For a modulating signal,

$$m(t) = 2 \cos(100t) + 18 \cos(2000\pi t)$$

(a) Write expressions for $y_{PM}(t)$ and $y_{FM}(t)$ (t) when carrier amplitude, $A = 10$, carrier frequency $\omega_c = 10^6$ rad/s, $k_f = 1000\pi$ and $k_p = 1$.
(b) Estimate the bandwidths of $y_{PM}(t)$ and $y_{FM}(t)$ (t).

P3.4 Consider the following FM signal:

$$v(t) = 5\cos\left[10^6 t + 3 \int_{-\infty}^{t} m(t)dt \right] V$$

(a) What is the frequency deviation of $v(t)$?
(b) Suppose max $|m(t)| = 4000\pi$, what is the maximum frequency deviation of $v(t)$?
(c) If $m(t)$ has a bandwidth of 5 kHz, what is the bandwidth of $v(t)$?

P3.5 The first-generation AMPS (American Mobile Phone Service) has the following specifications for FM transmission:

Maximum frequency deviation: 15 kHz
Message signal bandwidth: 5 kHz

(a) Determine the modulation index and bandwidth of the FM transmitter.

(b) For AMPS FM transmission, if the input SNR is 10 dB, determine the output SNR of the FM detector.

(c) If the input SNR is increased by 5 dB, what is the corresponding increase in the output SNR?

P3.6 A cellular signal has a psd of $S_x(f) = (|f| - 900\,\text{MHz})/60\,\text{kHz}$ and passes through a noisy channel with a constant psd of $S_n(f) = 10^{-3}$ W/Hz, for $|f| \leq 900$ MHz, and zero elsewhere. The noisy signal is then passed through a Wiener reconstruction filter, $H_o(f)$.

(a) Plot the psd of the input signal and noise, respectively.

(b) Calculate the input SNR, at the input of the filter.

(c) Calculate the SNR improvement at the output of the Wiener filter.

Computer Laboratory

In the computer laboratory, we will attempt to simulate the modulation systems described in the previous section, and get a better idea of the waveforms and spectra of the modulated signals.

C3.1 Simulation of AM system using Simulink® system block approach

Simulate the AM system as shown in Figure 3.16, using Simulink software. Obtain the power spectrum of the conventional AM signal (Double sideband with carrier, or DSB-C), which includes the

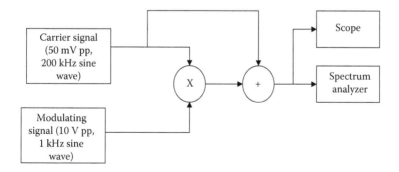

FIGURE 3.16
Block diagram of an AM system.

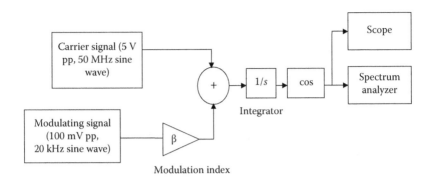

FIGURE 3.17
Block diagram of an FM system.

carrier signal. The power contained in the Fourier coefficients is given as follows:

$$Pc_n, comp = |c_n|^2 \, (\text{mW})$$

(a) Note down the frequency and amplitude of the carrier and two sidebands.

(b) Repeat part (a) by removing the link connecting the carrier to the adder in Figure 3.13. This is the double sideband suppressed carrier, or DSB-SC.

C3.2 Simulation of FM system using Simulink system block approach.
The simulation of FM basically involves the implementation of Equation 3.18. The process involves addition, integration, and cosine function to generate the frequency-modulated signal.

(a) Simulate the FM system as shown in Figure 3.17, using Simulink software.

(b) Obtain the time domain plot and power spectrum plot of the FM signal.

(c) Note down the frequency and amplitude of the carrier and first *four* sidebands.

Hardware Laboratory

H3.1 Fabrication and testing of AM transistor circuit
Two signal generators are used in this AM circuit shown in Figure 3.18, one representing a high-frequency (1 KHz) RF carrier,

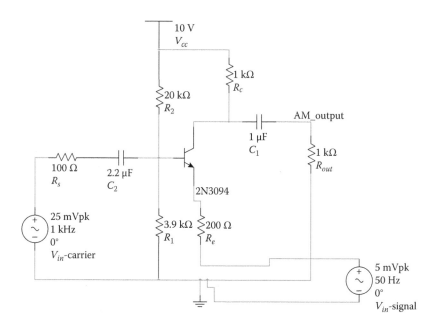

FIGURE 3.18
AM circuit.

and the other signal generator is used to inject a 50 Hz audio signal. The two signals are mixed and amplified by the transistor and an amplitude-modulated signal appears at the collector of the 2N3904. Fabricate the circuit shown in Figure 3.18 and measure the carrier signal, audio signal, and the modulated signal across the output resistor load. Write down the frequency and amplitude of each peak of the signals.

The signal you just recorded is the Double Sideband-Suppressed Carrier (DSB-SC). Next produce conventional AM (DSB-C) by adding carrier to the DSB-SC. Do this by connecting the carrier signal to the modulated signal output. Note down the carrier and sideband amplitudes and frequencies on the dynamic signal analyzer.

Comparison of simulated and measured data

(a) Compare the measured and computed power spectrum of the carrier and two sidebands in mW, after normalizing the peak values of the fundamental to 1 mW (0 dBm).

(b) Compute the % error between the computed and measured power spectrum (mW).

The % error is defined as

$$\% \, error = \frac{|Pc_n, comp - Pc_n, meas|}{Pc_n, comp}$$

(c) Repeat steps (a) and (b) using a DSB-C system, by adding the carrier signal.

H3.2 Fabrication and testing of FM transistor circuit

The simple FM circuit is shown in Figure 3.19 and is built using the general purpose 2N3904 transistor. The carrier frequency is determined by the tank or oscillator circuit, comprising of the inductor L_1 and variable capacitor V_{C1} connected in parallel. The frequency of oscillation is determined by the equation:

$$f_c = \frac{1}{2\pi\sqrt{L_1 * V_{C1}}}$$

Assuming $L_1 = 0.1$ μH and $C_1 = 50$ pF, we obtain $f_c \sim 50$ MHz. The variable capacitor can be tuned to get exactly 50 MHz as the unmodulated carrier. Fabricate the circuit shown in Figure 3.19 and measure the carrier signal, audio signal, and the modulated FM signal in both time and frequency domains. Note the frequencies and amplitudes of the carrier and first *four* sidebands.

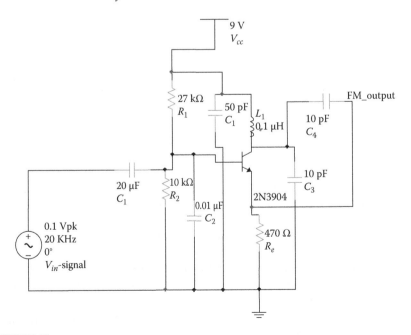

FIGURE 3.19
FM circuit.

Comparison of simulated and measured data

(a) Compare the measured and computed power spectrum of the carrier and four sidebands in mW, after normalizing the peak values of the fundamental to 1 mW (0 dBm).

(b) Compute the % error between the computed and measured power spectrum (mW).

The % error is defined as follows:

$$\% \, error = \frac{|Pc_n, comp - Pc_n, meas|}{Pc_n, comp}$$

4

Second-Generation Systems: Digital Modulation

Baseband signals such as speech, music and video are naturally occurring *analog* signals. *Digital* signals, on the other hand, are obtained by *coding* analog signals, and making it simpler to transmit the information. The *Morse code* is probably one of the oldest, and a classic example of digital coding; the entire English alphabet is represented by a combination of *dashes* and *dots*. For example, the emergency message SOS (Save Our Souls) is represented by the following code:

$$S \quad O \quad S$$
$$\cdots \quad \text{---} \quad \cdots$$

The main advantage is that even though the English language has 26 different looking and different sounding alphabets, all of them can be reduced to a combination of the same two characters, the dot and the dash. At the receiver, only these two characters need to be recognized correctly for reconstructing any message accurately. Of course, the receiver has the additional task of *decoding* the actual message from the combination of dots and dashes.

Hence, the processes of *analog-to-digital (A/D)* conversion at the transmitter and *digital-to-analog conversion (D/A)* at the receiver are integral sections of the entire communication system. Digital communications has proven to be a very efficient means of transporting speech, music, video, and data over different kinds of transmission media, such as satellite, microwave, fiber-optic, coaxial, and cellular channels. One special advantage that digital communication holds over analog communication is in the superior handling of *noise* in the channel. We will now discuss *Pulse Code Modulation (PCM)* in the next section, which is one of the basic forms of A/D systems.

Example

Consider creating a uniform code, using *circles* (o) and *crosses* (x) to represent the English alphabet. However, unlike the Morse code, this uniform code has the requirement that each alphabet have the same number of characters.

(a) How many characters would this code require to represent the entire English alphabet?
(b) How many characters would the code require if it were to represent the entire English alphabet and also the numbers 1–10?

Solution

(a) Since the entire English alphabet with 26 letters has to be represented with a uniform code of 2 symbols, we can use the following logic:
1 character can represent **2** levels: o (circle) and x (cross)
2 characters can represent **4** levels: oo ox xo xx
3 characters can represent **8** levels: ooo oox oxo oxx xoo xxo xox xxx
4 characters can represent **16** levels
5 characters can represent **32** levels
Since we require 26 levels, 5 characters will be sufficient to represent the entire English alphabet.

(b) Since the entire English alphabet, with 26 letters, and 10 numbers (totally 36 levels) have to be represented with a uniform code, we would require at least **6** *characters*, which will generate 64 levels. (*Hint: Try to obtain the general equation that relates the number of levels to the number of characters; this equation appears later in this chapter.*)

4.1 Pulse Code Modulation

PCM is one of the earliest developed methods of A/D conversion. The PCM process, as shown in Figure 4.1, converts an analog continuous-time signal such as speech or music into a digital binary bit stream. The three fundamental steps in the PCM process are *time sampling, amplitude quantization*, and *binary encoding*.

4.1.1 Time Sampling

The first step in the PCM process is *time sampling*, where the continuous-time signal $x(t)$, as shown in Figure 4.2a, is sampled uniformly at an interval of T s.

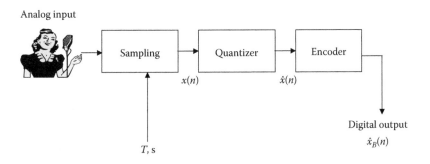

FIGURE 4.1
Block diagram of PCM.

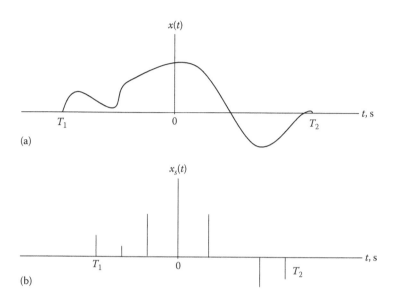

FIGURE 4.2
Sampling process. (a) Continuous-time signal and (b) sampled signal.

The output of the sampling process is the discrete-time signal, $x_s(t) = x(nT)$ or $x(n)$; $n = 0, 1, 2, ..., N - 1$, as shown in Figure 4.2b. Two important questions arise at this time:

- *What is the appropriate value of the sampling interval, T s, or inversely, what is the appropriate value of the sampling frequency $f_s = 1/T$ in cycles per second or Hertz (Hz)?*
- *Is it possible to recover x(t) "exactly" from the sample values x(n): n = 0, 1, 2, ..., N − 1?*

The answer to the first question is given by the *Nyquist sampling theorem,* which states: *If x(t) is a bandlimited signal with the maximum signal frequency* Ω_m, *rad./s then x(t) is uniquely determined from its samples x(n): n = 0, 1, 2, ..., N − 1, if the sampling interval* $T \leq \pi/\Omega_m$ *s, or, alternately, if the sampling frequency* $f_s \geq \Omega_m/\pi$ *Hz.* The threshold sampling frequency Ω_m/π Hz is called the *Nyquist sampling rate,* and the corresponding time threshold π/Ω_m is called the *Nyquist sampling interval.*

The answer to the second question is given by the interpolation formula in Equation 4.1. If the sampling satisfies the Nyquist theorem, give above then the recovered signal values (between the samples) are given by

$$x_r(t) = \sum_{n=0}^{N-1} x(n) \frac{\sin[\pi(t - nT)/T]}{\pi(t - nT)/T} \tag{4.1}$$

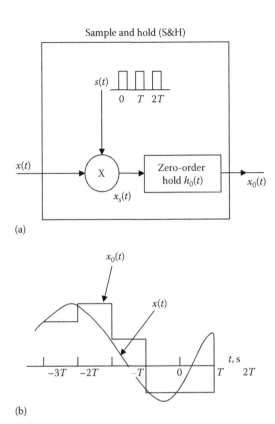

(a)

(b)

FIGURE 4.3
Practical sampling: (a) system (b) output.

However, *practical sampling* is different from the ideal sampling described in this section, and by Equation 4.1. One of the practical problems in ideal sampling is the impossibility of generating ideal impulses, with a zero time width. A practical sampling system would be like the *sample and hold* (S&H) circuit shown in Figure 4.3a, which would generate the sampled output, $x_0(t)$, shown in Figure 4.3b.

Example

Determine the Nyquist sampling rate and Nyquist sampling interval for each of the following analog signals:

(a) $1 + \cos(2000\pi t) + \sin(4000\pi t)$
(b) $\sin(2000\pi t)/(\pi t)$

Solution

(a) The maximum frequency contained in this signal is $\Omega_m = 4000\pi$ rad/s. Hence, the Nyquist rate is $\Omega_m/\pi = 4000$ Hz, and the Nyquist interval is 1/4000 or 0.00025 sec.

(b) The maximum frequency contained in this signal is $\Omega_m = 2000\pi$ rad/s, since the Fourier transform of sinc(at) = sin(at)/at function is a rectangular pulse in the interval ($-a$, a) rad/s. Hence, the Nyquist rate is $\Omega_m/\pi = 2000$ Hz, and the Nyquist interval is 1/2000 or 0.0005 sec.

4.1.2 Amplitude Quantization

The second stage in the A/D process is amplitude quantization, where the sampled discrete-time signal $x(n)$, $n = 0, 1, 2, ..., N - 1$ is quantized into a finite set of output levels $\hat{x}(n)$, $n = 0, 1, 2, ..., N - 1$. The quantized signal $\hat{x}(n)$ can only take one of L levels, which are designed to cover the dynamic range $-x_M \le x(n) \le x_M$, where x_M is the maximum amplitude of the signal. Both uniform and nonuniform quantizers will be considered in this section.

4.1.2.1 Uniform Quantizer

The design of an *L-level uniform quantizer* is detailed in a four-step process.

Step 1: Dynamic range of the signal

Fix the dynamic range of the sampled signal $-x_M \le x(n) \le x_M$.

Step 2: Step size of quantizer

The step size of the uniform quantizer is given as follows:

$$\text{Step size } \Delta = \frac{2x_M}{L} \tag{4.2}$$

The step size can be either integer or fraction and is determined by the number of levels L. For binary coding, L *is usually a power of* 2, and practical values are 256 (=2^8) or greater.

Step 3: Quantizer implementation

Draw the *input/output or staircase diagram* of the quantizer, as shown in Figure 4.4. The x-axis of the staircase diagram represents the input sampled signal $x(n)$, and the y-axis represents the quantized output $\hat{x}(n)$. As is seen from Figure 4.4a, the input levels are in integral multiples of $\Delta/2$, while the output levels are in integral multiples of Δ, *with output zero level included*. Such a quantizer is termed *midtread* quantizer, whereas a *midriser* quantizer, as shown in Figure 4.4b, does not include output zero level, and has the reverse structure of the midtread quantizer.

Step 4: Quantizer error and signal-to-noise ratio (SNR)

The quantizer error is calculated as follows:

$$e(n) = \hat{x}(n) - x(n), \quad n = 0, 1, 2, ..., N-1$$

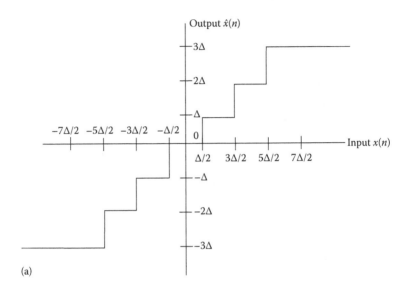

(a)

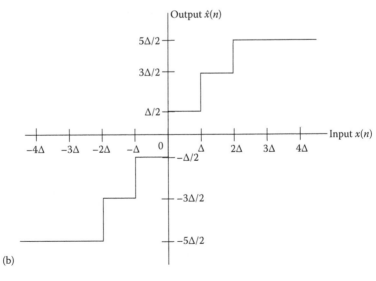

(b)

FIGURE 4.4
(a) Midtread and (b) midriser quantizers.

A figure of merit of the quantizer is defined by *the quantizer SNR* given as follows:

$$SNR = 10\log\left[\frac{\sigma_x^2}{\sigma_e^2}\right] \qquad (4.3)$$

In Equation 4.3, the *variance of the input signal* $x(n)$ is given as follows:

$$\sigma_x^2 = \overline{x^2(n)} - [\overline{x(n)}]^2$$

where $\overline{x^2(n)}$, the *mean squared value* of the input signal, is given by the following equation:

$$\overline{x^2(n)} = \frac{x^2(0) + x^2(1) + x^2(2) + \cdots + x^2(N-1)}{N} \qquad (4.4)$$

and the mean value of the input signal, $\overline{x(n)}$, is given by the following equation:

$$\overline{x(n)} = \frac{x(0) + x(1) + x(2) + \cdots + x(N-1)}{N} \qquad (4.5)$$

The variance of the quantization error is given by a simplified expression [4]:

$$\sigma_e^2 = \frac{\Delta^2}{12} \qquad (4.6)$$

Practical quantizers used for high-quality music work at *SNR* values around 90 dB. *Nonuniform quantizers* such as μ-law and *A*-law quantizers are widely used around the world to improve the *SNR* value. A detailed analysis of nonuniform quantizers is given in the next section.

4.1.2.2 Nonuniform Quantizer

The most important nonuniform quantization technique is logarithmic quantization (μ-*law* in United States, Japan, and Canada, and *A-law* in Europe, Africa, Asia, South America, and Australia), which has been used very successfully for speech digitization. This technique evolved from the fundamental property of speech, which has a Gamma or Laplacian probability density in amplitude, highly peaked at about zero value. However, even though low amplitudes of speech are more probable than large amplitudes, a uniform quantizer amplifies all signals equally.

The principle behind non-uniform quantization is to preprocess (compress) the sampled signal before it enters the uniform quantizer, such that the processed signal occupies the full dynamic range of the quantizer. However,

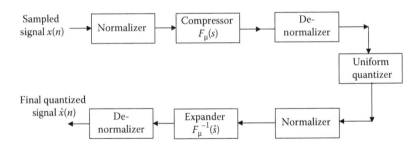

FIGURE 4.5
Block diagram of nonuniform quantization.

the output of the uniform quantizer has to be postprocessed (expanded) to extract the true quantized signal. This dual process is called *logarithmic companding*, which is a combination of *compression* and *expanding*. The non-uniform quantization process is explained in a series of *four steps*, which is also illustrated in Figure 4.5.

Step 1: Dynamic range and normalizing the sampled signal

Fix the dynamic range of the sampled signal $-x_M \leq x(n) \leq x_M$. Normalize the sampled signal $x(n)$, by its peak amplitude x_M, to yield the normalized signal $s(n)$:

$$s(n) = \frac{x(n)}{x_M}$$

with a dynamic range $-1 \leq s(n) \leq 1$.

Step 2: Signal compression

This step basically preprocesses the sampled signal, to provide more amplification for the lower amplitude samples, and less amplification for the higher amplitude samples. The compression function is as follows:

$$F_\mu(s) = \frac{\ln(1 + \mu|s|)}{\ln(1 + \mu)} \operatorname{sgn}(s) \tag{4.7}$$

where
  $s(n)$ is the normalized sampled signal and
  μ is the compression parameter, usually taken as 255

The *ln* function is the natural logarithm to base e, and sgn(s) is the *signum* function defined as follows:

$$\operatorname{sgn}(s) = 1, \quad s > 0$$
$$= -1, \quad s < 0$$

Step 3: Processing by uniform quantizer

The compressed output, $F_\mu(s)$, is input to a uniform L-level quantizer, which has been described in detail in the previous section.

Step 4: Signal expanding

The output of the uniform quantizer, $\hat{s}(n)$, is passed through the inverse expanding function, in order to resynthesize the input signal. The expanding function is as follows:

$$F_\mu^{-1}(\hat{s}) = \frac{1}{\mu}[(1+\mu)^{|\hat{s}|}-1]\operatorname{sgn}(\hat{s}) \tag{4.8}$$

where $-1 \leq \hat{s}(n) \leq 1$.

Step 5: Signal denormalization

The final step in the non-uniform quantization process is the denormalization of the signal, $\hat{s}(n)$, to yield the final quantized signal:

$$\hat{x}(n) = \hat{s}(n)x_M$$

Example

A sampled signal that varies between −2 and 2 V is quantized using B bits. What value of B will ensure an rms quantization error of less than 5 mV?

Solution

The quantization error variance $\sigma_e^2 = \Delta^2/12 = (2x_M/L)^2/12$. Given that $x_M = 2$ V and variance $= (5\text{ mV})^2$, we obtain the number of quantization levels $L = 230.94$. Since $L = 2^p$, where q is the number of bits/sample, we finally obtain $p = 7.85$, which can be rounded to the next higher integer level of 8 bits.

Example

Consider the ramp signal $x(t) = 2t$, over the range of $t = (0, 1)$ s. For a sampling interval of 0.1 s, obtain the sampled signal, quantized signal, error signal, and quantizer SNR, assuming a uniform four level quantizer.

Solution

The sampling time intervals are as follows: $t = [0\ 0.1\ 0.2\ 0.3\ 0.4\ 0.5\ 0.6\ 0.7\ 0.8\ 0.9\ 1.0]$.

(a) Then, the corresponding signal samples are as follows:

$$x(n) = [0\ 0.2\ 0.4\ 0.6\ 0.8\ 1.0\ 1.2\ 1.4\ 1.6\ 1.8\ 2.0]$$

(b) The quantizer step size $\Delta = 2x_M/L = (2 \times 2)/4 = 1.0$, and the four quantizer output levels (for a midtread quantizer) are

$$-1.0, 0.0, 1.0 \text{ and } 2.0.$$

Since a quantizer with even number of levels cannot have a symmetrical number of levels for positive and negative values, the question arises whether we should take the extra level on the positive side or the negative side. Either choice can be made, since the number of quantizer levels is usually very high, and no appreciable difference in quantizer error will be observed.

(c) The quantizer staircase diagram is shown in Figure 4.6. Following the diagram, the quantized output is

$$\hat{x}(n) = \begin{bmatrix} 0 & 0 & 0 & 1.0 & 1.0 & 1.0 & 1.0 & 1.0 & 2.0 & 2.0 & 2.0 \end{bmatrix}$$

(d) The error vector is: $e(n) = \hat{x}(n) - x(n) = [0\ -0.2\ -0.4\ 0.4\ 0.2\ 0\ -0.2\ -0.4\ 0.4\ 0.2\ 0]$. It is important to note that any element of the error vector should not have a magnitude greater than or equal to $\Delta/2$ or 0.5 in this case.

(e) The quantizer SNR is given by Equation 4.3:

$$SNR = 10\log\left[\frac{\sigma_x^2}{\sigma_e^2}\right]$$

where

$$\sigma_x^2 = \overline{x^2(n)} - \overline{[x(n)]}^2$$

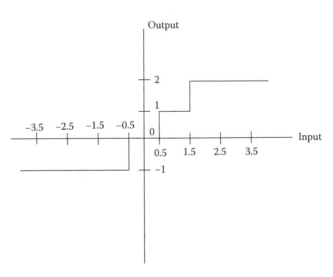

FIGURE 4.6
Quantizer staircase diagram.

where $\overline{x^2(n)}$, the mean squared value of the input signal, is given by

$$\overline{x^2(n)} = \frac{\begin{array}{l} 0^2 + 0.2^2 + 0.4^2 + 0.6^2 + 0.8^2 + 1.0^2 + 1.2^2 + 1.4^2 \\ + 1.6^2 + 1.8^2 + 2.0^2 \end{array}}{11} = 1.4$$

and the mean value of the input signal $\overline{x(n)}$ is given by

$$\overline{x(n)} = \frac{0 + 0.2 + 0.4 + 0.6 + 0.8 + 1.0 + 1.2 + 1.4 + 1.6 + 1.8 + 2.0}{11} = 1$$

The variance of the quantization error is obtained from Equation 4.6:

$$\sigma_e^2 = \frac{\Delta^2}{12} = \frac{1}{12} = 0.0833$$

Hence, the variance $\sigma_x^2 = 1.4 - 1^2 = 0.4$, and the quantizer SNR is

$$SNR = 10 \log \left[\frac{0.4}{0.0833} \right] = 6.8141 \, \text{dB}$$

This is a very low SNR specification; in practical quantizers, the number of quantizer levels is usually very high (e.g., $L = 256$ is common) and SNR can be improved significantly.

4.1.3 Digital Encoding

The *third and last stage* in the A/D process is digital encoding, where the quantized signal $\hat{x}(n)$, $n = 0, 1, 2, ..., N - 1$ is encoded to yield the final *digital signal* $\hat{x}_B(n)$, $n = 0, 1, 2, ..., N - 1$. As an illustration, if the number of quantizer levels $L = 8 = 2^3$, the number of binary bits required to encode all the L levels is 3. Table 4.1 illustrates the encoding procedure, using *two's complement coding*, which is very convenient in the decoding process at the receiver.

The two's complement code (TCC) is quite easily obtained from the offset binary code (OBC), by *complementing the leftmost bit of the OBC*. The decimal form of the TCC includes *both positive and negative numbers* and is given by the following equation:

$$\text{Decimal number} = -a_0 2^0 + a_1 2^{-1} a_2 2^{-2} + \cdots + a_B 2^{-B} \tag{4.9}$$

where the original binary number is $[a_0 \; a_1 \; a_2 \; a_B ... a_B]$. The decoding process at the receiver is illustrated in Table 4.2.

TABLE 4.1

Binary Encoding Process

Quantizer Level (for $L = 8$)	Offset Binary Code (3-bit)	Two's Complement Code (3-bit)
−4	000	100
−3	001	101
−2	010	110
−1	011	111
0	100	000
1	101	001
2	110	010
3	111	011

TABLE 4.2

Binary Decoding Process

TCC (3-bit)	Decimal Value (from Equation 4.9)	Actual Quantizer Level
100	−1	−4
101	−3/4	−3
110	−1/2	−2
111	−1/4	−1
000	0	0
001	1/4	1
010	1/2	2
011	3/4	3

From Table 4.2, it is seen that the recovered quantized value from the binary bit stream can be obtained easily as follows:

Quantized value = Decimal value of TCC × Peak value of sample value (x_M)

In practical communication systems, non-binary systems are also used to increase system capacity. Examples are *quadrature* systems that use *four* states and *octal* systems that used *eight* states instead of the two states used in binary systems. As a comparison, the number of levels possible in these different systems can be calculated as follows, for a given number of digits, q:

Binary: Number of encoding levels $L = 2^q$

Quadrature: Number of encoding levels $L = 4^q$

Octal: Number of encoding levels $L = 8^q$

As can be seen, the number of encoding levels increases exponentially as the coding changes from binary to octal.

4.1.4 Transmission Rate and Shannon's Maximum Capacity Theorem

The final transmission rate or *bit rate* of the system is calculated as follows:

$$\text{Bit rate } R_b = \text{Sampling rate} \times \text{Number of bits/symbol}$$

$$= f_s \times p \text{ bps} \tag{4.10}$$

The unit of bit rate is bits/second or bps. Bit rate of the system is directly related, for example, to the Internet speed which governs how fast we can view online pages, or how fast we can upload and download files. As early as in 1949, *Claude Shannon* gave his landmark theorem, which related the maximum bit rate possible, or *channel capacity*, to the available bandwidth as follows:

$$\text{Channel capacity } C = B \log_2(1 + SNR) \tag{4.11}$$

where
  B is the bandwidth of the transmission channel
  SNR is the signal-to-noise ratio

Example

A television signal (audio and video) has a bandwidth of 4.5 MHz. The signal is sampled, quantized, and binary-coded to obtain a PCM signal.

(a) Determine the sampling rate if the signal has to be sampled at a rate 20% above the Nyquist rate.
(b) If the samples are quantized into 1024 levels, determine the number of binary pulses required to encode each sample.
(c) Determine the binary pulse rate (bits per second) of the binary-coded signal, and the minimum bandwidth required to transmit this signal.
(d) What would be the channel bandwidth required to transmit at this bit rate, assuming a channel SNR of 20 dB?

Solution

(a) Nyquist rate = 2 × Maximum signal frequency = 2 × 4.5 MHz = 9 MHz.

   Sampling rate = 1.2 × 9 MHz = 10.8 MHz.

(b) Number of levels = 1024 = 2^p => Number of binary pulses/ sample = p = 10.

(c) Binary pulse rate = Sampling rate (Hz) × Number of pulses/ sample

   = 10.8 MHz × 10 = 108 Mbps

(d) The relation between channel bandwidth and bit rate is given by Shannon's capacity theorem:

   Channel capacity C (or maximum possible bit rate)
     = $B \log_2(1 + SNR)$

Putting in the given values: $C = 108$ Mbps and $SNR = 20$ dB $=>$
$10^{20/10} = 100$, we obtain:

$$108 \times 10^6 = B \log_2(1 + 100)$$

$$=> \text{Bandwidth } B = 108 \times 10^6/\log_2(101)$$

$$= 108 \times 10^6/6.658 = 16.22 \text{ MHz.}$$

Note: $\log_2(x) = \log_{10}(x)/\log_{10}(2)$

4.1.5 Line Coding and Pulse Shaping Technique

Once we get the 1s and 0s from the PCM process that was outlined in the previous sections, the next step is to convert this information into electrical voltages or pulses. A simple solution would be assign a rectangular pulse, as seen earlier in Figure 2.3b, with levels of 0 or 5 V to represent the digital states of 0 and 1, respectively. Such a pulse structure is called *unipolar*, whereas pulse levels of ±5 V to represent 0 and 1 is called *bipolar*, an example of which is the *Manchester code*. Hence, we would expect a stream of pulse voltages, corresponding to the binary data, with a pulse period $T_b = 1/R_b$, where R_b is the bit rate as defined earlier. A symbol is a collection of bits, in which case we would define the symbol period $T_s = 1/R_s$, where R_s is the symbol rate.

However, while a rectangular pulse is geometrically elegant and symmetrical, its spectrum, as seen from Figure 2.3b, has infinite bandwidth, primarily due to the sharp rise and fall nature of the pulse waveform. When such a pulse passes through a bandlimited channel, its spectrum will be truncated, and Fourier analysis shows that a narrowing of spectrum will cause the pulse to spread in time. This pulse spreading results in adjacent pulses interfering with one another, and this distortion is termed as *inter symbol interference (ISI)*. The solution to this problem was given by *Nyquist*, who identified the need for *pulse shaping* of the original rectangular pulse to satisfy the mathematical criterion that bears his name.

Practically, the latter pulse shaping is achieved by passing the default rectangular pulse through a pulse-shaping filter, like the popular *raised cosine roll-off filter*. The filter transfer function and its impulse response are given in Equations 4.12 and 4.13.

$$H_{RC}(f) = \begin{cases} 1, & 0 \le |f| \le \dfrac{1-r}{2T_s} \\ 0.5\left\{1+\cos\left[\dfrac{\pi(2|f|T_s-1+r)}{2r}\right]\right\}, & \dfrac{1-r}{2T_s} \le |f| \le \dfrac{1+r}{2T_s} \\ 0, & |f| > \dfrac{1+r}{2T_s} \end{cases} \quad (4.12)$$

$$h_{RC}(t) = \frac{\sin\left(\pi t/T_s\right)}{\pi t} \frac{\cos\left(\pi rt/T_s\right)}{1-\left(4rt/2T_s\right)^2} \tag{4.13}$$

Finally, the maximum symbol rate possible through a baseband raised cosine filter is given by Equation 4.14 below:

$$R_s = \frac{1}{T_s} = \frac{2B}{1+r} \tag{4.14}$$

where B is the filter bandwidth.

Example

In a digital modulation system with number of encoding levels, $L = 16$, calculate the following:

(a) Determine the minimum transmission bandwidth required to transmit data at a rate of 20 kbps with zero ISI.
(b) Determine the transmission bandwidth if Nyquist criterion pulses with a roll-off factor $r = 0.5$ are used to transmit data.

Solution

(a) Using Equation 4.13 with $R_s = 20$ kbps, and $r = 0$ (zero ISI), we obtain

$$B = \frac{R_s}{2} = 10 \text{ kHz}$$

(b) Again, using Equation 4.13 with $R_s = 20$ kbps, and $r = 0.5$, we obtain

$$B = \frac{R_s\left(1+r\right)}{2} = \frac{20\left(1.5\right)}{2} = 15 \text{ kHz}$$

4.2 Digital Modulation Systems

In the previous section, we have discussed the process of A/D conversion, which results in the analog signal being converted into a coded combination of 1s and 0s. Once the coding is completed, as shown in Section 4.1.3, the 1s and 0s need to be converted into *voltage pulses* for transmission. As in the case of analog modulation schemes that were discussed in Chapter 3, a strong sinusoidal signal is still the base carrier for these voltage pulses, and digital modulation can be implemented by using any of the schemes described.

4.2.1 Digital AM or Phase Shift Keying

Mathematically, the concept of digital amplitude modulation is quite simple: Assume a carrier signal $A \cos(\omega_c t)$, which is switched between two levels, corresponding to the two binary levels, 0 and 1:

$$0 => A \cos(\omega_c t)$$

$$1 => -A \cos(\omega_c t)$$

This can also be viewed as the carrier undergoing a phase shift process as follows:

$$0 => A \cos(\omega_c t + 0°)$$

$$1 => A \cos(\omega_c t + 180°)$$

Since there are only two levels of modulation, this scheme is called as *Binary Phase Shift Keying (BPSK)*. The two phase shift angles, 0° and 180°, can be generated by dividing a full circular span of 360° into two (binary) parts to yield a spacing of 180°. A BPSK transmitter/receiver system is shown in Figure 4.7. As in analog AM systems, the BPSK signal is also obtained by multiplication of the carrier signal and the digital pulse data, which is the result of the A/D conversion that was detailed in Section 4.1.

Similarly, this principle can be extended to quadrature and octal systems too. In *Quadrature Phase Shift Keying (QPSK)*, the circle is divided into four

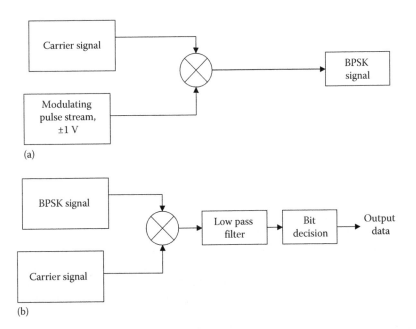

FIGURE 4.7
Block diagram of BPSK system: (a) transmitter (b) receiver.

(quadrature) parts to yield a spacing of 90°. This gives four phase shift values 0°, 90°, 180°, and 270° and the carrier values as follows:

$$0 => A \cos(\omega_c t + 0°) = A \cos(\omega_c t)$$

$$1 => A \cos(\omega_c t + 90°) = -A \sin(\omega_c t)$$

$$2 => A \cos(\omega_c t + 180°) = -A \cos(\omega_c t)$$

$$3 => A \cos(\omega_c t + 270°) = A \sin(\omega_c t)$$

A QPSK transmitter/receiver system is shown in Figure 4.8. At the input of the *transmitter*, the digital data's even bits (i.e., bits 0, 2, 4, and so on) are split and are multiplied with a carrier to generate the *in-phase(I)* signal. At the same time, the data's odd bits (i.e., bits 1, 3, 5 and so on) are split from the data stream and are multiplied with the same carrier (with a 90° phase shift) to generate the *quadrature phase(Q)* signal. The two BPSK signals are then added

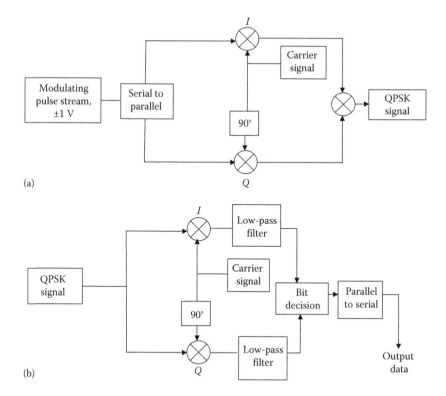

FIGURE 4.8
Block diagram of QPSK system: (a) transmitter (b) receiver.

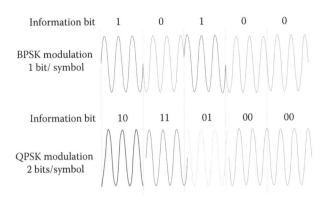

FIGURE 4.9
Carrier modulation in BPSK and QPSK systems.

together for transmission and, since they have the same carrier frequency, they occupy the same portion of the radio-frequency spectrum.

Likewise, at the *receiver*, the QPSK signal is multiplied twice, once with the carrier, and again with the 90° phase-shifted carrier to generate the even and odd bit data streams. These data streams are detected and combined in the parallel to serial block to yield the final received data. Typical BPSK and QPSK signals are shown in Figure 4.9; note the abrupt phase change, corresponding to the changing digital level. In the QPSK signal, the four digital levels 0, 1, 2, and 3 are converted to the 2-bit sequence format of 00, 01, 10, and 11.

In *Octal Phase Shift Keying (OPSK)*, the circle is divided into eight (octal) parts to yield a spacing of 45°. This gives eight phase shift values 0°, 45°, 90°, 135°, 180°, 225°, 270°, and 315° and the carrier values as follows:

$$0 => A \cos(\omega_c t + 0°)$$

$$1 => A \cos(\omega_c t + 45°)$$

$$2 => A \cos(\omega_c t + 90°)$$

$$3 => A \cos(\omega_c t + 135°)$$

$$4 => A \cos(\omega_c t + 180°)$$

$$5 => A \cos(\omega_c t + 225°)$$

$$6 => A \cos(\omega_c t + 270°)$$

$$7 => A \cos(\omega_c t + 315°)$$

4.2.2 Digital FM or Frequency Shift Keying

Similar to analog FM systems, digital FM systems also modulate the carrier frequency, but only between *two values*, as shown below:

$$0 \Rightarrow A \cos[(\omega_c + \Delta\omega)t)$$

$$1 \Rightarrow A \cos[(\omega_c - \Delta\omega)t)$$

A typical binary frequency shift keying (BFSK) signal is shown in Figure 4.10; note the abrupt frequency change, corresponding to the changing digital level. An FSK transmitter/receiver system is shown in Figure 4.11. At the *transmitter*, the voltage-controlled oscillator (VCO) is tuned by the input data stream to switch between two carrier frequencies f_1 and f_2. In the *receiver*, the FSK signal is mixed with both carrier frequency signals to generate the bit stream data. The data are averaged by integrating over each bit period, T, and then a decision block (> or < 0) finalizes the process.

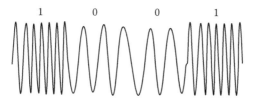

FIGURE 4.10
Carrier modulation in FSK system.

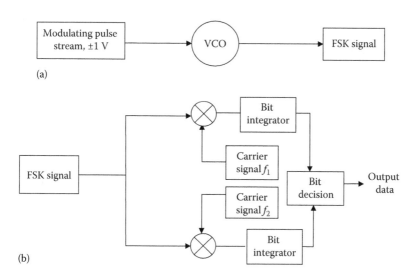

FIGURE 4.11
Block diagram of FSK system: (a) transmitter (b) receiver.

4.2.3 Differential Phase Shift Keying

Differential phase shift keying (DPSK) is a different form of phase modulation that conveys data by changing the phase of the carrier wave. This implies that the phase of the carrier changes every time there is a change of data, but the phase of the carrier remains same as long as there is no transition. Figure 4.12 shows a typical DPSK signal; note that every time the input data changes from 0 to 1 or 1 to 0, the phase changes, otherwise remaining constant.

A DPSK transmitter/receiver system is shown in Figure 4.13. In the transmitter, the input data stream is fed back with a unit delay (of one sampling period T) and is added to the initial data stream, as shown in Figure 4.13a, where the symbol $\oplus$ denotes *modulo-2 addition*. The added output is BPSK modulated to yield the final DPSK signal. Likewise, in the *receiver*, as shown in Figure 4.13b, the DPSK signal is BPSK demodulated and then subtracted

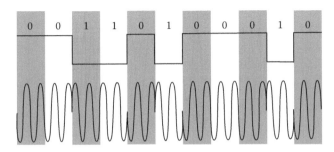

FIGURE 4.12
Carrier modulation in DPSK system.

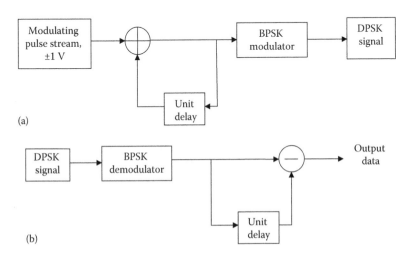

FIGURE 4.13
Block diagram of DPSK system: (a) transmitter (b) receiver.

from its unit delayed version, using *modulo-2 subtraction*. Both modulo-2 addition and subtraction can be achieved by the XOR circuit, with the following logic:

Adding: $0 + 0 = 0$ $0 + 1 = 1$ $1 + 0 = 1$ $1 + 1 = 0$
Subtracting: $0 + 0 = 0$ $0 + 1 = 1$ $1 + 0 = 1$ $1 + 1 = 0$

4.3 BER and Bandwidth Performance in Digital Modulation Systems

In digital systems, bits that are transmitted across a channel can be corrupted with noise, and a 1 at the transmitter could appear as a 0 at the receiver, or vice versa. While there are *error-correcting* mechanisms, as will be described in the next section, a few error bits could still get through the channel. The error performance is measured by the *bit error rate (BER)*, which is a unitless quantity, represented as a percentage. A BER of 0.03 or 3% implies that, on an average, 3 bits of every 100 transmitted bits would be in error. Practical communication systems have very small BERs in the order of 10^{-5} or 10^{-6}.

The equations for BER, or *error probability*, P_e, for the digital communication schemes discussed earlier are given in Equation 4.14. In these equations, the term E_b/N_0 represents *SNR*, where E_b is the *signal energy/bit* (Unit: Joules), and N_0 is the *noise spectral density* (Unit: Watts/Hz or Joules). Hence E_b/N_0 is a dimensionless quantity and is often represented in dB: $\left[E_b/N_0\right]_{dB} = 20\log_{10}\left[E_b/N_0\right]$.

$$P_{e,BPSK} = Q\left[\sqrt{\frac{2E_b}{N_0}}\right]$$

$$P_{e,FSK} = Q\left[\sqrt{\frac{1.217E_b}{N_0}}\right] \tag{4.15}$$

$$P_{e,DPSK} = Q\left[\sqrt{\frac{E_b}{N_0}}\right]$$

The Q function is given by the integral:

$$Q(z) = \int_{z}^{\infty} e^{-x^2/2}dx \tag{4.16}$$

The easiest method to estimate the Q function approximately is through the chart shown in Table 4.3.

TABLE 4.3

Q Function Table

Z	Q(z)	z	Q(z)
0.0	0.50000	2.0	0.02275
0.1	0.46017	2.1	0.01786
0.2	0.42074	2.2	0.01390
0.3	0.38200	2.3	0.01072
0.4	0.34458	2.4	0.00820
0.5	0.30854	2.5	0.00621
0.6	0.27425	2.6	0.00466
0.7	0.24196	2.7	0.00347
0.8	0.21186	2.8	0.00256
0.9	0.18406	2.9	0.00187
1.0	0.15866	3.0	0.00135
1.1	0.13567	3.1	0.00097
1.2	0.11507	3.2	0.00069
1.3	0.09680	3.3	0.00018
1.4	0.08076	3.4	0.00034
1.5	0.06681	3.5	0.00023
1.6	0.05480	3.6	0.00016
1.7	0.04457	3.7	0.00011
1.8	0.03593	3.8	0.00007
1.9	0.02872	3.9	0.00005

For values of z greater than 3.9, the following approximation can be used:

For $z > 3.9$;

$$Q(z) \sim \frac{1}{z\sqrt{2\pi}} e^{-z^2/2}$$

(4.17)

Bandwidth of the digital modulation schemes are given in the following equation, in terms of the bit rate R_b.

$$B_{BPSK} = R_b$$

$$B_{FSK} = R_b$$

(4.18)

$$B_{DPSK} = \frac{R_b}{2}$$

An interesting fact, from eqn. 4.18, is that DPSK systems can have *twice* the bit rate of BPSK and FSK systems, with the same available bandwidth.

Example

An additive white Gaussian noise (AWGN) channel has a noise density of $N_0 = -10$ dBm/Hz. Various digital modulation schemes are compared with the same bit energy level, $E_b = 0$ dBm.

(a) Compare probability of error P_e for the following systems: BPSK, FSK, and DPSK.
(b) Compare the bandwidth of the different systems for a bit rate of 1 Mbps.

Solution

(a) SNR $= E_b/N_0$

$E_b = 0$ dBm => 1 mW

$N_0 = -10$ dBm/Hz = 0.1 mW

Hence, $E_b/N_0 = 1/0.1 = 10$.

$$P_{e,BPSK} = Q\left[\sqrt{\frac{2E_b}{N_0}}\right] = Q\left[\sqrt{20}\right] = Q[4.47] = 3.91 \times 10^{-6}$$

$$P_{e,FSK} = Q\left[\sqrt{\frac{1.217E_b}{N_0}}\right] = Q\left[\sqrt{12.18}\right] = Q[3.49] = 2.42 \times 10^{-4}$$

$$P_{e,DPSK} = Q\left[\sqrt{\frac{E_b}{N_0}}\right] = Q\left[\sqrt{10}\right] = Q[3.16] = 7.89 \times 10^{-4}$$

(b) $B_{BPSK} = R_b = 1$ MHz.

$B_{FSK} = R_b = 1$ MHz.

$B_{DPSK} = R_b/2 = 0.5$ MHz.

4.3.1 Noise Correction and Filtering in Digital Modulation Systems

The very nature of digital systems makes them very efficient, since the entire communication is based on the two basic digital units, 0 and 1. Even though the transmitted bits may be subject to noise in the channel, each bit has to be recovered between just two choices at the receiver; hence, there is always a 50% chance of getting it right! Some important techniques and systems that contribute to the success and wide usage of digital communication systems will be described in the next sections.

4.3.1.1 Error-Detecting Codes

Error-correcting codes (ECCs) are systems that add an extra bit of information in the code to check for a preset arrangement, such as *parity*. To explain, let us consider two sample 5-bit messages as shown in the following:

Message A: 0 1 1 1 0

Message B: 1 1 0 0 0

To introduce *even parity*, for example, we will add an extra 1 or 0, which makes the number of 1s as *even*. Using this logic, the parity-coded messages become

Message A: 0 1 1 1 0 **1**
Message B: 1 1 0 0 0 **0**

If we consider the case of these messages getting corrupted by the channel, and arriving at the receiver with one of the bits, let us say, the first bit, in error, then the corrupted messages would look as follows:

Corrupted message A: 1 1 1 1 0 **1**
Corrupted message B: 0 1 0 0 0 **0**

Then, the receiver would detect the error in both messages, since the number of 1s in both messages is *odd*. However, the receiver may not be able to detect multiple bit errors, if the number of 1s in the corrupted message is still even. Other forms of EDCs include *cyclic redundancy checks*, *checksums*, and *cryptographic hash functions*.

4.3.1.2 Error-Correcting Codes

EDCs, which were explained in the previous section, can detect errors but cannot correct them. An extension of the EDC would be if the receiver requests the transmitter, through a reverse channel, to resend the message in which the error was detected. ECCs, which do not require a reverse channel, include *convolutional codes*, and *block codes*, such as *Reed–Solomon* and *BCG (Bose–Chaudhuri–Hocquenghem)* codes. *Turbo codes* and *low-density parity-check codes (LDPC)* are newer codes with high efficiency of error correction.

4.3.2 Equalization and Channel Compensation

Equalization is probably the most powerful and widely applicable method to limit channel distortion such as noise and fading. To understand the equalization process, let us recall the Wiener filter that was discussed in Chapter 3. If the channel is stationary, then the filter transfer function $H_o(f)$ is designed using the average power spectral densities (psd) of the signal and noise. This filter can then clean up the noise in the signal and increase SNR.

However, the important question is: What happens if the channel is *not stationary*? This implies that we would require a *changing* filter transfer function, based on the channel transfer function at any particular moment of time. Such a process is called *adaptive*, or as the name implies, the filter adapts with the channel. There are *two* fundamental problems in this quest for an adaptive filter: *first*, how do we determine the instantaneous transfer function of the channel, and *secondly*, how do we design a filter that can keep adapting to the changing channel? The first problem is solved by sending a *training pulse* at regular intervals of time along with the data, as shown in the following:

Training pulse–Data–Training pulse–Data–Training pulse–Data....

The training pulse is a fixed sequence $p(t)$, which gets modified in the channel with impulse response $h_c(t)$, and is received as $r(t)$. Writing this relation in the frequency domain, we have the received training pulse spectrum as

$$R(\omega) = P(\omega)\, H_c(\omega) \tag{4.19}$$

Since the training pulse is a *known sequence*, unlike the data which is random, we can determine the channel frequency response, $H_c(\omega)$, using Equation 4.19. Then, the data, which immediately follows the training pulse, can be compensated for the channel by using a *compensating filter, H(\omega)*, with the transfer function:

$$H(\omega) = \frac{1}{H_c(\omega)} \tag{4.20}$$

Such an inverse operation is called an *equalizer,* since the filter effectively cancels out the effect of the channel on the data.

Additionally, we see that the training pulse is sent periodically to keep measuring the channel transfer function, $H_c(\omega)$, which could be changing with time. As the channel changes, the recovery filter, $H(\omega)$, would also have to change to compensate the data accurately. This process is called *tracking* and requires an *adaptive digital filter,* which can change its frequency response instantaneously with the channel. Since the channel is typically an *analog* system, the corresponding analog equalizer filter, $H(\omega)$, is given by Equation 4.19. The impulse response of the equalizer filter, $h(t)$, can be sampled at a time interval $t = nT$ to yield the digital impulse response, $h(nT)$ or $h(n)$. A typical adaptive digital filter structure is shown in Figure 4.14.

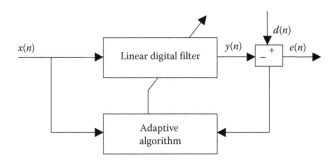

FIGURE 4.14
Adaptive digital filter.

The input–output relation of the filter is given, as discussed in Chapter 2, in Equation 2.11, as follows:

$$y(n) = \sum_{m=0}^{N-1} x(m)h((n-m)_N), \quad 0 \le n \le N-1$$

which can be rewritten as follows:

$$y(n) = \sum_{m=0}^{N-1} h(m)x((n-m)_N), \quad 0 \le n \le N-1 \tag{4.21}$$

The digital filter is defined by the coefficients, $h(n)$, $0 \le n \le N - 1$, and the filter coefficients are continuously changed using the adaptive algorithm, as shown in Figure 4.14. The adaptive algorithm aims to *minimize* the difference error, $e(n)$, which is defined as the difference between the channel input and the filter output:

$$e(n) = d(n) - y(n) \tag{4.22}$$

Such a filter can be realized in hardware by using a *microprocessor,* which can instantaneously modify the filter coefficients. The following example will illustrate the filter reconstruction process.

Example

Consider a *time-varying* channel having the following frequency response:

$$H_c(f) = \begin{cases} \dfrac{\omega^2 + 20,000}{8}, & \text{during daytime} \\[2ex] \dfrac{\omega^2 + 40,000}{8}, & \text{during nighttime} \end{cases}$$

(a) Design an adaptive *analog* filter to compensate for both states of the channel. Determine the impulse response in both states.
(b) Design an adaptive *digital* filter to compensate for both states of the channel. Determine the impulse response in both states, assuming a sampling frequency of 40 kHz.

Hint: Use the following Fourier transform pair:

$$e^{-a|t|} \iff \frac{2a}{a^2 + \omega^2}$$

Solution

(a) The *analog* equalizer filter frequency response is obtained from Equation 4.17 as

$$H(f) = \frac{1}{H_c(f)} = \begin{cases} \dfrac{8}{\omega^2 + 20{,}000}, & \text{during daytime} \\[3mm] \dfrac{8}{\omega^2 + 40{,}000}, & \text{during nighttime} \end{cases}$$

Using the given Fourier transform pair, we can obtain the equalizer impulse response as follows:

$$h(t) = \begin{cases} \dfrac{8}{2\sqrt{20{,}000}}\, e^{-\sqrt{20{,}000}|t|}, & \text{during daytime} \\[4mm] \dfrac{8}{2\sqrt{40{,}000}}\, e^{-\sqrt{40{,}000}|t|}, & \text{during nighttime} \end{cases}$$

(b) The digital filter impulse response, $h(n)$, is obtained by sampling the analog filter impulse response, $h(t)$ at $t = nT$, where T is the sampling interval. Using a sampling frequency of 40 kHz, and a corresponding sampling interval $T = 1/40$ kHz $= 25$ μs, we can write down the impulse response of the digital equalizer filter as follows:

$$h(n) = \begin{cases} \dfrac{8}{2\sqrt{20{,}000}}\, e^{-\sqrt{20{,}000}\,(0.000025)|n|}, & \text{during daytime} \\[4mm] \dfrac{8}{2\sqrt{40{,}000}}\, e^{-\sqrt{40{,}000}\,(0.000025)|n|}, & \text{during nighttime} \end{cases}$$

and simplifying, the final form of the digital equalizer impulse response is

$$h(n) = \begin{cases} 0.028 e^{-0.0035|n|}, & \text{during daytime} \\[2mm] 0.02 e^{-0.0050|n|}, & \text{during nighttime} \end{cases}$$

Problem Solving

P4.1 Design a uniform eight-level quantizer for an input signal with a dynamic range of ±10 V.

(a) Calculate the quantization error vector for an input signal of $x(n) =$ [−4.8 −2.4 2.4 4.8].

(b) Calculate the quantization error for the same input signal if the quantizer is preceded by a $\mu = 255$ compander (compressor/expander).

P4.2   A continuous time signal, $x(t)$, has a bandwidth of 10 kHz, is sampled at a rate of 22 kHz, and uniformly quantized and encoded at 8 bits/sample. The signal is properly scaled so that $|x(n)| < 128$ for all n.

(a) Determine the variance of the quantization error σ_e^2.

(b) If the sampling rate was increased by 16 times, how many bits per sample would you use to maintain the same level of quantization error?

P4.3   A voice signal $m(t)$, which is band-limited to 3,200 Hz, multiplies the function $\cos(\omega_c t)$ where $\omega_c = 2\pi(10,000)$ rad/s.

(a) Assuming ideal sampling, specify a sampling rate such that the signal $x(t) = m(t) \cos(\omega_c t)$ can be exactly recovered using an ideal band-pass filter.

(b) What are the ideal band-pass filter cut-off frequencies?

P4.4   A sampled signal that varies between −4 and 4 V is quantized using B bits.

(a) What value of B will ensure an rms quantization error of less than 5 mV?

(b) Consider the ramp signal $x(t) = 2t$, over the range of $t = (0, 1)$. For a sampling interval of 0.1 s, obtain the sampled signal, quantized signal, error signal, and SNR, assuming a uniform 8-level quantizer.

P4.5   A compact disk (CD) records audio signals digitally using a modulation scheme. Assume the audio signal bandwidth to be 15 kHz.

(a) What is the Nyquist sampling rate?

(b) If the Nyquist samples are quantized into 65,536 levels, and then binary-coded, determine the number of binary digits required to encode a sample.

(c) Determine the number of bits/second required to encode the signal.

P4.6   Design a raised cosine roll-off filter for a GSM system with symbol rate of 270.833 kHz and roll-off factor of 0.5.

(a) What is the maximum filter bandwidth possible for this roll-off factor?

(b) Approximately plot the impulse response and frequency response of the filter.

P4.7 An AWGN channel has a noise density of $N_0 = -15$ dBm/Hz. Various digital modulation schemes are compared with the same bit energy level, $E_b = 0$ dBm.

(a) Compare probability of error P_e for the following systems: BPSK, FSK, and DPSK.

(b) Compare the bandwidth of the different systems for a bit rate of 2 Mbps.

Computer Laboratory

C4.1 Practical circuit for A/D and D/A conversion using Simulink®
The schematic of a practical A/D circuit is shown in Figure 4.15a and the schematic of a practical D/A circuit is shown in Figure 4.15b. The ideal S&H circuit, as shown in Figure 4.15a, is equivalent to impulse train

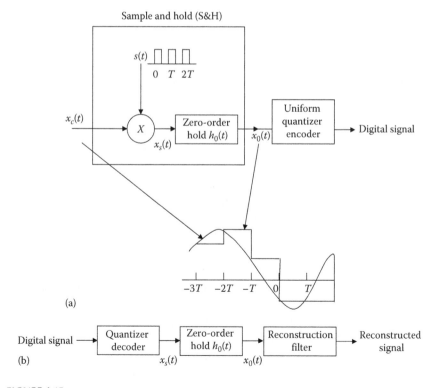

FIGURE 4.15
(a) A/D and (b) D/A systems.

modulation followed by linear filtering with the zero-order hold (ZOH) system. The output of the ZOH system is the staircase waveform shown in Figure 4.15a. The sample values are held constant during the sampling period of T, s. In the design of the S&H circuit on Simulink, the following three important blocks will have to be designed accurately:

The source signal block $x_s(t)$: Since these block parameters are fixed, no further design is necessary on this block.

The pulse train block $s(t)$: Two important parameters will have to be designed for this block. The first is the pulse amplitude, and the pulse period T s. You can assume a rectangular pulse with 50% duty cycle (i.e., half period on, and half period off). Since this pulse train samples the source signal, its frequency should be many times higher than that of the source signal.

The zero-order hold block: One important parameter will have to be designed for this block, which is the sampling period of the hold circuit. The sampling period of the circuit should be sufficient to hold the sample value over each period of the pulse train.

(a) Select an appropriate audio signal from the Simulink DSP blockset library as the test signal in this simulation. Plot the signal on the scope and the FFT scope to obtain the frequency content of the signal. This will provide information on the maximum frequency content of the signal, and then require sampling rate limits. Sample the signal at the *Nyquist rate.*

(b) Design the required parameters of the A/D circuit given in Figure 4.15a to obtain the sampled signal. Plot the output of the zero-order hold circuit as seen on the scope block of the Simulink program.

(c) Design a uniform quantizer to convert the sampled signal into quantized signal output in numerical or binary form.

(d) Design the required parameters of the D/A circuit given in Figure 4.15b to reconstruct the signal at the receiver. The reconstruction filter is modeled as a low-pass filter (analog or digital) with cutoff frequency as the sampling frequency utilized in the A/D process.

(e) Plot the reconstructed signal, and compare with transmitted analog signal, and determine the error signal.

(f) Repeat the entire simulation for a case of *undersampling*: choose a sampling frequency smaller than the Nyquist rate (e.g., half the Nyquist rate), and plot the transmitted signal, reconstructed signal, and the error signal.

(g) Repeat the entire simulation for a case of *oversampling*: choose a sampling frequency larger than the Nyquist rate (e.g., twice the Nyquist rate), and plot the transmitted signal, reconstructed signal, and the error signal.

C4.2 Repeat Problem C4.1, however, with the following modifications:

(a) Introduce a μ-*law compressor* before the uniform quantizer, as shown in Figure 4.5. Similarly introduce a μ-*law expander* after the uniform quantizer. Assume μ = 255.

(b) As in Problem C4.1, plot the transmitted signal, reconstructed signal, and the error signal for the cases of *undersampling, oversampling,* and *Nyquist* sampling.

(c) Compare the error in reconstruction, between the cases of uniform quantization and nonuniform quantization.

C4.3 Simulate the *differential pulse code modulation (DPCM)* system, shown in Figure 4.16, using Simulink.

Transmitter

Assume an input signal: $s(t) = 10 \sin(5\pi t) + 5 \sin(8\pi t)$. In this simulation, one period or multiple periods of the signal can be processed. The input analog signal $s(t)$ is sampled at a rate much higher that the Nyquist rate (~25–50 times). This generates very closely spaced samples $s(nT)$, which have a very great degree of correlation between adjacent values. In traditional PCM, the signal $s(nT)$ is directly quantized and encoded. However, in DPCM, the following difference is quantized:

$$e(t) = s(nT) - s((n-1)T)$$

The difference signal is quantized as follows:

$$\hat{e}(t) = \delta \quad \text{if } e(t) > 0$$

$$\hat{e}(t) = -\delta \quad \text{if } e(t) < 0$$

$$\hat{e}(t) = 0 \quad \text{if } e(t) = 0$$

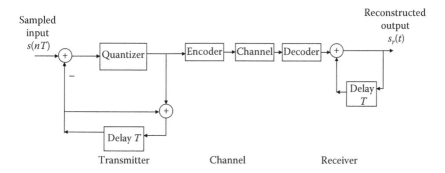

FIGURE 4.16
Block diagram of DPCM system.

or compactly as $\hat{e}(t) = \text{sgn}[e(t)]$, where sgn is the signum function and the step size should be selected to satisfy the condition: $\delta \ll |s(t)|_{max}$. Thus, the final signal $\hat{e}(t)$ consists of pulses, with amplitude $\pm\delta$. Plot the quantized signal $\hat{e}(t)$ for at least one period of the original signal $s(t)$.

Channel
Model the channel as a system gain of 1.0.

Receiver
The receiver consists of an integrator, which sums the pulses $e(n)$, and generates the reconstructed signal $s_r(t)$.

(a) Plot the input and reconstructed signals on the same graph, and determine the mean-squared error between them.

(b) Plot the error signal between the input and the reconstructed signals.

Comment on the advantages that DPCM would have over the PCM system that was discussed in Section 4.1.

C4.4 Develop a Simulink model to implement the BPSK transmitter/ receiver system shown in Figure 4.7, using Table 4.4, which identifies blocks from the Simulink library to represent system components:

(a) Plot the BPSK output on the scope and verify that it corresponds to the input pulse stream.

(b) Obtain the bandwidth of the BPSK signal from the spectrum analyzer display.

C4.5 Develop a Simulink model to implement the QPSK transmitter/ receiver system shown in Figure 4.8, using Table 4.5, which identifies blocks from the Simulink library to represent system components:

(a) Plot the QPSK output on the scope and verify that it corresponds to the input pulse stream.

(b) Obtain the bandwidth of the QPSK signal from the spectrum analyzer display.

TABLE 4.4

BPSK System Components and Equivalent Simulink® Blocks

BPSK System Component	Corresponding Simulink Block
Carrier signal	Sine wave at 900 MHz
Modulating pulse stream ±1 V	Pulse generator at 10 MHz
Channel	Gain (−70 dB)
Low-pass filter	Analog filter design (low pass)
Bit decision	Threshold detector

Note: Scale down frequencies by a factor of 10 to reduce simulation time.

TABLE 4.5

QPSK System Components and Equivalent Simulink® Blocks

QPSK System Component	Corresponding Simulink Block
Carrier signal	Sine wave at 900 MHz
Modulating pulse stream ±1 V	Pulse generator at 10 MHz
Serial to parallel	Serial to parallel
90°	Phase shifter
I, Q	Multipliers
Channel	Gain (−70 dB)
Low-pass filter	Analog filter design (low pass)
Bit decision	Threshold detector
Parallel to serial	Parallel to serial

Note: Scale down frequencies by a factor of 10 to reduce simulation time.

TABLE 4.6

FSK System Components and Equivalent Simulink® Blocks

FSK System Component	Corresponding Simulink Block
Modulating pulse stream ±1 V	Pulse generator at 1 MHz
VCO	VCO
Channel	Gain (−70 dB)
Carrier signals	Sine wave sources
Bit integrator	Discrete time integrator
Bit decision	Threshold detector

Note: Use basic circuits shown to replicate VCO and Detector if Simulink blocksets are not available. For VCO, use Figure 3.14 with the signal input replaced by pulse input. Scale down frequencies by a factor of 10 to reduce simulation time.

C4.6 Develop a Simulink model to implement the FSK transmitter/receiver system shown in Figure 4.11, using Table 4.6, which identifies blocks from the Simulink library to represent system components:

(a) Plot the FSK output on the scope and verify that it corresponds to the input pulse stream.

(b) Obtain the bandwidth of the FSK signal from the spectrum analyzer display.

C4.7 Develop a Simulink model to implement the DPSK transmitter/receiver system shown in Figure 4.13, using Table 4.7, which identifies blocks from the Simulink library to represent system components:

(a) Plot the DPSK output on the scope and verify that it corresponds to the input pulse stream.

(b) Obtain the bandwidth of the BPSK signal from the spectrum analyzer display.

TABLE 4.7

DPSK System Components and Equivalent Simulink® Blocks

DPSK System Component	Corresponding Simulink Block
Modulating pulse stream ±1 V	Pulse generator at 1 MHz
Unit delay	Unit delay
Modulo-2 adder	Modulo-2 adder
BPSK modulator	Use blocks from Problem C4.4
Channel	Gain (–70 dB)
BPSK demodulator	Use blocks from Problem C4.4
Modulo-2 subtractor	Modulo-2 subtractor

Note: Scale down frequencies by a factor of 10 to reduce simulation time.

Hardware Laboratory

H4.1 Design and construction of a simple S&H circuit

The S&H circuit using an FET switch is shown in Figure 4.17, and can sample rapidly changing voltages that arise from the input signal $x_c(t)$. The op-amp acts as a high-input impedance voltage follower. When a pulse train $s(t)$ is high at the sample input, the FET is turned on (during the "on" cycle) and acts as low resistance to the input signal. When the sample pulse is absent, the FET is turned off and acts as high impedance. The desired voltage is held by capacitor C_1, which is isolated from the output by the high-input impedance op-amp. When the switch is closed, the capacitor charges to $x_{c(max)}$. After the switch is opened, the capacitor remains charged and $x_0(t)$ will be at the same

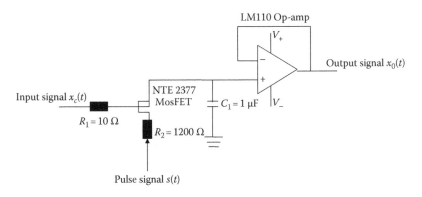

FIGURE 4.17
S&H circuit.

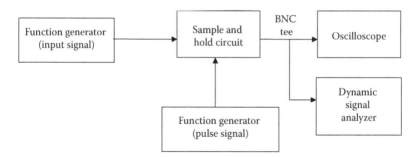

FIGURE 4.18
Measurement setup for S&H circuit.

potential as the capacitor. The sampled voltage will be held temporarily, with the time determined by leakage in the circuit.

(a) Connect the circuit as shown in Figure 4.18 and apply a 1 kHz (input signal frequency) sinusoidal signal to the input of the S&H circuit. Use a 10 kHz (sampling frequency) pulse signal to drive the sample input of the S&H circuit. Observe the sampled output at the output of the circuit on an oscilloscope.

(b) Repeat the experiment for the maximum possible input signal frequency. Please note that the sample frequency should be accordingly increased in order to obtain the required number of samples.

(c) Plot the spectra of the input and output signals of the S&H circuit on the Keysight 35665A Dynamic Signal Analyzer. Comment on the differences between the two spectra.

H4.2 Design and construction of a simple BPSK circuit
Fabricate the circuit shown in Figure 4.19, which is identical to the AM circuit in Figure 3.18, with the exception that we now have a *pulse input*.

(a) Plot the measured BPSK output on the scope.

(b) Obtain the bandwidth of the BPSK signal by connecting the output to a spectrum analyzer.

Comparison of simulated and measured data

(a) Compare the measured and simulated power spectrum of the BPSK signal.

(b) Compute the % error between the computed and measured BPSK signal bandwidth.

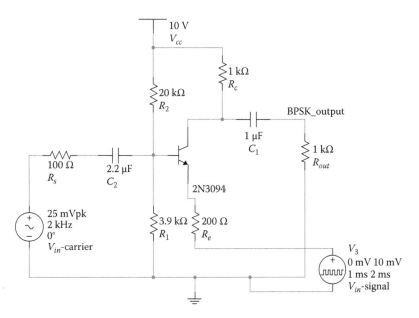

FIGURE 4.19
Practical BPSK circuit.

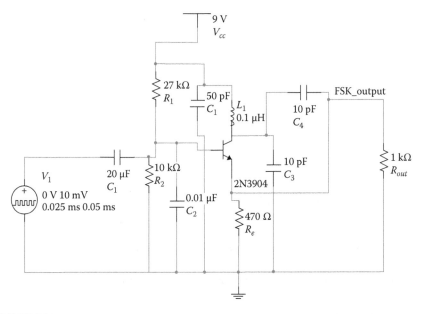

FIGURE 4.20
Practical FSK circuit.

H4.3 Design and construction of a simple FSK circuit

Fabricate the circuit shown in Figure 4.20, which is identical to the AM circuit in Figure 3.19, with the exception that we now have a *pulse input*.

(a) Plot the measured FSK output on the scope.
(b) Obtain the bandwidth of the FSK signal by connecting the output to a spectrum analyzer.

Comparison of simulated and measured data

(a) Compare the measured and simulated power spectrum of the FSK signal.
(b) Compute the % error between the computed and measured FSK signal bandwidth.

5

Third-Generation Systems: Wideband Digital Modulation

Second-generation communication systems ushered in the era of digital technology, which created very compact data sets with limited states that increased the efficiency and speed of signal transmission. This improved performance was made possible with techniques that could only be realized by digital hardware, like *equalization* which significantly improved noise reduction, and *Time Division Multiplexing* (*TDM*), which increased the capacity and number of available user channels.

However, a key component was still missing from transmission systems: *high bandwidth*. First- and second-generation systems could offer a maximum bandwidth of 50 kHz/user, which was good enough for audio but remarkably low for video and high-speed Internet usage, which require at least 1 MHz of bandwidth. Thus, the stage was set for a new technology called *spread spectrum* to supply the much needed megahertz bandwidth.

5.1 Principle of Spread Spectrum Communications

First- and second-generation communication systems work on the principle of *frequency division*; for example, if your local cellular system had a total bandwidth of 50 MHz and 1000 channels, then each channel gets a bandwidth of 50 MHz/1000 = 50 kHz. If we allocate one channel per user, then each user gets a maximum bandwidth of 50 kHz.

Let us say each user wants the *entire bandwidth* of 50 MHz to obtain higher Internet speed. Is it even possible? One solution then is that all the users *share* the total 50 MHz, rather than divide the bandwidth. This is similar to, for example, many city dwellers sharing a large park. The only problem is that since all users will be sharing the common bandwidth, they will have total interference, and all cell phone conversations or data would be totally public. The solution to this interference problem was devised by use of *pseudo-noise*, or *PN codes*, whose application will be explained in the following example.

To understand the concept of PN codes, let us consider three users, A, B, and C, who want to share a *common broadband channel*, and would like to send one data each on the *same* channel while maintaining zero interference between themselves. The example data that the users want to send are as follows:

User A data: 12
User B data: 7
User C data: −10

The first step is to assign an individual PN code to each of the users, which is a collection of *positive and negative ones*, as shown below

A: [1 1 1 1]
B: [1 −1 1 −1]
C: [1 1 −1 −1]

This is an example of a set of PN codes, whose special properties are listed in the following:

- The *length* of the code, N, is the length of the vector; $N = 4$ in this case. Practical systems may have much larger code lengths, for example, the 3G system developed by Qualcomm has a length of 128!
- These code vectors are *orthogonal*, which means that the dot product of any two code vectors is zero. For example,

$$A \cdot B = [1 \quad 1 \quad 1 \quad 1] \cdot [1 \quad -1 \quad 1 \quad -1]$$
$$= 1(1) + 1(-1) + 1(1) + 1(-1)$$
$$= 1 - 1 + 1 - 1 = 0$$

Similarly, it can be shown that $B \cdot C = C \cdot A = 0$. The next two steps will show how messages are coded and decoded using the orthogonality property of PN sequences.

Transmission coding:

User 1: User 1 Data × User 1 code = 12 × [1 1 1 1] = [12 12 12 12]
User 2: User 2 Data × User 2 code = 7 × [1 −1 1 −1] = [7 −7 7 −7]
User 3: User 3 Data × User 3 code = −10 × [1 1 −1 −1] = [−10 −10 10 10]

Transmitted data = Sum of user data

$$= [12 \quad 12 \quad 12 \quad 12] + [7 \quad -7 \quad 7 \quad -7] + [-10 \quad -10 \quad 10 \quad 10]$$
$$= [9 \quad -5 \quad 29 \quad 15]$$

This combined data is sent on the single broadband channel to be decoded at the receiver.

Receiver decoding:
In order to recover each user data from the combined transmitted data, we need to combine it with the user PN codes as follows:

Received User 1 data = Combined data·User 1 code/Code Length

$$= [9 \quad -5 \quad 29 \quad 15] \cdot [1 \quad 1 \quad 1 \quad 1]/4$$

$$= 48/4 = 12 \ (User \ 1 \ data \ recovered \ successfully!)$$

Similarly,

Received User 2 data = Combined data·User 2 code/Code Length

$$= [9 \quad -5 \quad 29 \quad 15] \cdot [1 \quad -1 \quad 1 \quad -1]/4$$

$$= 28/4 = 7 \ (User \ 2 \ data \ recovered \ successfully!)$$

Received User 3 data = Combined data·User 3 code/Code Length

$$= [9 \quad -5 \quad 29 \quad 15] \cdot [1 \quad 1 \quad -1 \quad -1]/4$$

$$= -40/4 = -10 \ (User \ 3 \ data \ recovered \ successfully!)$$

Of course, we have not considered the effects of noise and fading which may affect the accuracy of data recovery, and increase the BER of the systems. These factors will be considered in the next sections, which describe the two main types of spread spectrum systems: *Frequency Hopping Spread Spectrum* (*FHSS*) and *Direct Sequence Spread Spectrum* (*DSSS*). In an FHSS sequence, each value in the PN sequence is known as a *channel number* and the inverse of its period as the *hop rate*. In a DSSS system, each bit in the PN sequence is known as a *chip* and the inverse of its period as *chip rate*.

5.2 Frequency-Hopping Spread Spectrum

Spread spectrum systems evolved during World War II, primarily as a technique for naval ships to avoid detection by enemy submarines. Since the submarine would track a ship by detecting its transmitting signal frequency, the approach adopted was to *keep changing the radio signal frequency.* For example, if a ship was transmitting radio signals at a constant frequency of 100 MHz, then it would be relatively easy for an enemy to detect and lock on to the ship. However, if the ship were transmitting at two or more frequencies, for example, 100, 150, and 200 MHz, then it would be more difficult to keep track of it.

This approach called *frequency hopping*, used initially for military applications, was later applied to the systems that we use now, such as Wi-Fi and Bluetooth, and will be explained in more detail in the upcoming sections. In modern wireless applications, users also hop through different frequencies; however, the hopping sequence should be different for different users to avoid interference. For example, consider three users; to avoid interference, the following hopping sequences are acceptable:

User A: (100, 150, 200) MHz

User B: (150, 200, 100) MHz

User C: (200, 100, 150) MHz

Note that the three users use *nonoverlapping* sequences. Hence, all three users share the large bandwidth of 100–200 MHz, without interfering with each other. The generation of such nonoverlapping sequences is possible by the use of digitally generated *PN codes*, which were discussed in the previous section.

5.2.1 FHSS Transmission and Reception

First- and second-generation communication systems are primarily a *one-stage modulation* process, with the message signal (analog or digital) modulating the carrier signal. While first- and second-generation systems only have the carrier modulation stage, third-generation spread spectrum systems have an additional modulation stage, where the *PN coding* is added to carrier-modulated signal before transmission. Likewise in the receiver, spread spectrum systems have a *two-stage demodulation* too. This aspect is depicted in Figure 5.1, which shows the main blocks in an FHSS transceiver (transmitter/receiver). Hence, the sequence of FHSS transmission and reception is as follows:

```
Input Data => FSK Modulation => PN Modulation =>
                    Channel =>
PN Demodulation => FSK Demodulation => Output Data
```

The *first stage of modulation* is typical frequency shift keying (FSK) carrier modulation, which was described in Chapter 4. The key difference between second-generation digital systems and spread spectrum systems comes in the *second stage of modulation*. The PN code generator, as shown in Figure 5.1, modulates the frequency synthesizer, which hops the carrier frequency through the *hop set*, or complete set of available carrier frequencies.

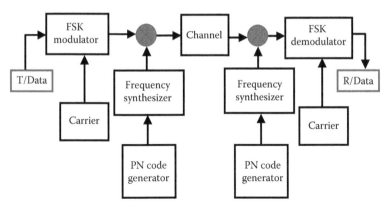

FIGURE 5.1
Block diagram of FHSS transmission and reception.

CF_1	CF_2	CF_3	CF_4	CF_5	CF_6	CF_7	CF_8

FIGURE 5.2
FHSS hopping sequence for User 1.

Figure 5.2 shows the hopping sequence for any one user: CF_1, CF_2, …, CF_8, assuming a hop set of *eight* carrier frequencies (CFs).

Practical FHSS systems, such as Bluetooth, have a much larger number of hopping channels. Note that each user in the FHSS system would hop through these same frequencies, however, in a *different sequence*. For example, User 2 might hop through the hop set in the following sequence shown in Figure 5.3. Comparing Figures 5.2 and 5.3, we see that for the same hop set position, the two carrier frequencies are always different, to avoid *interference* between two users. However, if such a situation does occur when two carrier frequencies are the same at a particular hop set position, it is called a *hit*. The *processing gain* (P_G) of the system is defined as the total number of hopping channels, and can be calculated, with reference to Figure 5.2 or 5.3, as follows:

$$P_G = \frac{W}{B} \tag{5.1}$$

where
B is the bandwidth of each hopping frequency
W is the bandwidth of the entire hop set

CF_8	CF_7	CF_6	CF_5	CF_4	CF_3	CF_2	CF_1

FIGURE 5.3
FHSS hopping sequence for User 2.

5.2.2 FHSS Bandwidth and BER Performance

As we have mentioned earlier, third-generation communication systems have a much increased bandwidth per user by sharing the entire spectrum available. The *bandwidth* of FHSS systems is given in terms of the baseband signal bandwidth as below:

$$BW_{FHSS} = P_G B_{FSK} \tag{5.2}$$

where
   B_{FSK} is the bandwidth of the FSK signal
   P_G is the processing gain given by Equation 5.2

The *bit error rate (BER)* performance of the FHSS system is given by the following equation:

$$P_e = 0.5 e^{-E_b/2N_0} \left(1 - \frac{M-1}{P_G} \right) + 0.5 \frac{M-1}{P_G} \tag{5.3}$$

where
   E_b/N_0 is the channel signal-to-noise ratio
   M is the number of users

In FHSS applications like *Bluetooth*, M could refer to the number of devices connected by the network, like printers, keyboards, headphones, and similar devices.

5.3 Direct-Sequence Spread Spectrum

Similar to FHSS modulation systems that were discussed in the earlier section, DSSS transmitting systems also have an additional modulation stage, where the *PN code* adds the second layer of modulation to the already carrier-modulated signal. Likewise in the receiver, DSSS systems have *two-stage demodulation* too. This aspect is depicted in Figure 5.4, which shows the main blocks in a DSSS transmitter and receiver.

5.3.1 DSSS Transmission and Reception

The principle involved in DSSS generation is quite different, though much simpler than the complex frequency hopping process. Figure 5.5a shows the input data sequence, with period $T = 1/Data\ rate$ (bps), and Figure 5.5b shows the PN sequence. The data is random; however, the PN sequence is a fixed, repeating sequence for a given user. For example, in this case, the PN

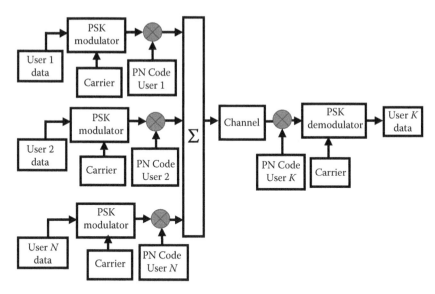

FIGURE 5.4
Block diagram of DSSS transmission and reception.

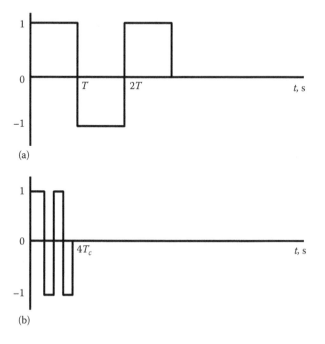

FIGURE 5.5
(a) Data sequence and (b) PN sequence for a user.

sequence is [1 –1 1 –1] with a length of 4, and to distinguish from data bits, PN pulses are called as *chips*. P_G of the DSSS system is defined as the ratio of the bit period to the chip period:

$$P_G = \frac{T}{T_c} \tag{5.4}$$

which is also equal to the length of the PN sequence.

5.3.2 DSSS Bandwidth and BER Performance

The *bandwidth* of the DSSS system is given in terms of the baseband signal bandwidth as follows

$$BW_{DSSS} = P_G \, B_{PSK} \tag{5.5}$$

where
  B_{PSK} is the bandwidth of the PSK signal
  P_G is the processing gain given by Equation 5.4

The BER performance of the DSSS system is given in Equation 5.6:

$$P_e = Q\left[\frac{1}{\sqrt{\dfrac{M-1}{3P_G} + \dfrac{N_0}{E_b}}} \right] \tag{5.6}$$

where
  E_b/N_0 is the channel signal-to-noise ratio
  M is the number of users
  $Q(x)$ is the Q function defined earlier in Chapter 4.

Example

(a) In the DSSS spread spectrum system, the incoming data rate is 15 kbps and the chip rate is 1.92 Mcps. It has been determined that for acceptable performance, that is, BER < 10^{-5}, the signal-to-noise ratio, E_b/N_0, must be greater than 20 dB. What is the maximum number of users that can be supported under these conditions?

(b) Repeat section (a) for a FHSS system, and calculate the maximum number of users possible.

Solution

(a) For a DSSS system:

$$P_e = Q \left[\cfrac{1}{\sqrt{\cfrac{M-1}{3P_G} + \cfrac{N_0}{E_b}}} \right] = 10^{-5} \text{(given)}$$

Hence, taking the inverse of the Q function:

$$\cfrac{1}{\sqrt{\cfrac{M-1}{3P_G} + \cfrac{N_0}{E_b}}} = 4.3$$

Putting processing gain $P_G = 1.92 \times 10^6 / 15 \times 10^3 = 128$

$E_b / N_0 = 20 \, \text{dB} => 100$

Number of users $M = 17$ users.

(b) For an FHSS system

$$P_e = 0.5 e^{-E_b/2N_0} \left(1 - \frac{M-1}{P_G} \right) + 0.5 \frac{M-1}{P_G} = 10^{-5} \text{(given)}$$

Putting processing gain $M = 128$

$$\frac{E_b}{N_0} = 20 \, \text{dB} => 100$$

$$= P_e = 0.5 e^{-50} \left(1 - \frac{M-1}{128} \right) + 0.5 \frac{M-1}{128} = 10^{-5}$$

$$\Rightarrow M = 1 \text{ user.}$$

5.4 Advantages and Disadvantages of Spread Spectrum Systems

The advantages include the following:

- *Noise immunity*: Due to the processing gain and digital processing a spread-spectrum-based system is more immune to interference and noise, as compared with first-and second-generation systems. This greatly reduces consumer electronic device–induced static noise that is commonly experienced by conventional analog wireless system users.

- *Multipath fading immunity:* Because of its inherent wide spectrum spread, a spread spectrum system is much less susceptible to multipath fading.

- *Wireless security:* In a spread spectrum system, a PN sequence is used to either modulate the signal in the time domain (DSSS) or generate the carrier frequency (FHSS). Due to the pseudorandom nature of the PN sequence, a spread spectrum system provides signal security that is not available to conventional analog wireless systems.

- *Longer operating range:* A spread spectrum device operated in the Industrial, Scientific and Medical (ISM) frequency band (~2.4 GHz) is allowed to have higher transmitted power due to its noninterfering nature; hence, the operating distance of such a device can be significantly longer than that of a traditional 1G or 2G wireless communication device.

- *Immunity from interception:* Since PN codes are long and orthogonal, it is very difficult for a random system to intercept and decode the signal; hence, interception is very hard.

- *Immunity from jamming:* It may seem that spread spectrum systems would be most susceptible to interference or jamming, due to the large bandwidth. However, if interference or jamming enters the spread spectrum system, the less are their effects on the system, because the power density of the signal after processing is lower, owing to the large signal bandwidth.

The following are the *challenges* in spread spectrum systems:

- *Increased bandwidth:* Since spread spectrum systems have inherently high bandwidth, this relates to requirement of wideband circuitry, and channel models for wideband systems.

- *Increased complexity:* Spread spectrum systems are two-stage modulation systems; hence, they are inherently more complex in design than first- and second-generation communication systems.

Problem Solving

P5.1   An FHSS system uses 50 kHz channels over a continuous 20 MHz spectrum, and 2 hops occur for each bit. Determine:

    (a)   The probability of error for a user operating at $E_b/N_0 = 15$ dB with 25 other FHSS users, which are independently frequency hopped.

(b) The probability of error for a user operating at $E_b/N_0 = 25$ dB with 150 other FHSS users, which are independently frequency hopped.

P5.2 In a BPSK-modulated DSSS system, the processing gain (Chip rate/ Bit rate) offered is 1000.

(a) If a *BER* of 10^{-6} is required, how many users can share the system? Assume that the receiver noise is negligible.

(b) Find the number of users for the system in (a) if a FHSS system is used. Assume FSK modulation, and same values *BER* and noise.

P5.3 A DSSS system has a 1.2288 Mchips per second (Mcps) code clock and a 9.6 kbps information rate.

(a) Calculate the processing gain.

(b) How much improvement information rate is achieved if the code generation rate is changed to 5 Mcps and the processing gain to 256?

Computer Laboratory

C5.1 Create a Simulink® model for the FHSS system shown in Figure 5.1, using Table 5.1, which identifies blocks from the Simulink library to represent system components:

TABLE 5.1

FHSS Components and Equivalent Simulink® Blocks

FHSS System Component	Corresponding Simulink Block
T/data	Pulse generator at 1 MHz bit rate
Carrier	Sine wave at 900 MHz
FSK modulator	VCO (voltage-controlled oscillator)
PN code generator	Pulse generator at 10 MHz
Frequency synthesizer	VCO
Channel	Gain block with 20 dB loss
FSK demodulator	PLL

Notes: Use basic circuits to replicate VCO and PLL if Simulink block-sets are not available. For VCO, use Figure 3.17 with the signal input replaced by pulse input; for PLL, use Figure 5.6. Scale down frequencies by a factor of 10 to reduce simulation time.

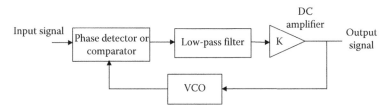

FIGURE 5.6
Block diagram of PLL.

A phase-locked loop (PLL), as shown in Figure 5.6, is a feedback control system that generates an output signal whose phase is related to the phase of the input *reference* signal. This circuit compares the phase of the input signal with the phase of the signal derived from its output oscillator and adjusts the frequency of its oscillator to keep the phases matched. The different signal from the phase detector is used to control the oscillator in a feedback loop.

C5.2 Create a Simulink model for the DSSS system shown in Figure 5.4, using Table 5.2, which identifies blocks from the Simulink library to represent system components:

C5.3 The PN sequence generator block, shown in Figure 5.7, generates a sequence of pseudorandom binary numbers using a linear-feedback shift register (LFSR). This block implements LFSR using a simple shift register generator (SSRG, or Fibonacci) configuration. The shift register is described by the *generator polynomial*, which is a primitive binary polynomial in z, $g_r z^r + g_{r-1} z^{r-1} + g_{r-2} z^{r-2} + \cdots + g_0$. The generator polynomial can be generated in either of the following formats:

(a) A vector that lists the coefficients of the polynomial in descending order of powers. The first and last entries must be 1. Note that the length of this vector is one more than the degree of the generator polynomial.

TABLE 5.2

DSSS Components and Equivalent Simulink® Blocks

DSSS System Component	Corresponding Simulink Block
T/data	Pulse generator at 1 MHz bit rate
Carrier	Sine wave at 900 MHz
BPSK modulator	Multiplier
PN code generator	Pulse generator at 10 MHz or use system from Figure 5.7
Channel	Gain block with 20 dB loss
BPSK demodulator	Multiplier

Notes: Use basic circuits to replicate VCO and PLL if Simulink blocksets are not available. For VCO, use Figure 3.17 with the signal input replaced by pulse input. Scale down frequencies by a factor of 10 to reduce simulation time.

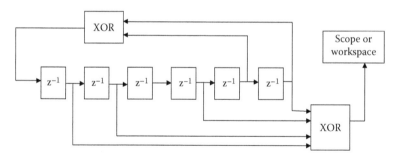

FIGURE 5.7
Block diagram of PN sequence generator.

(b) A vector containing the exponents of z for the nonzero terms of the polynomial in descending order of powers. The last entry must be 0.

For example, [1 0 0 0 0 0 1 0 1] and [8 2 0] represent the *same polynomial*, $p(z) = z^8 + z^2 + 1$. Referring to Figure 5.7, for the candidate generator polynomial, $p(z) = z^6 + z + 1$, the model generates a PN sequence of period 63, using the corresponding Simulink blocks. You can experiment with different initial states on the XOR blocks, by changing the value of *initial states* prior to running the simulation. Plot the PN outputs in each case.

6

Capacity of Communication Systems and Higher Generations

The evolution of communication systems from 1G to 3G interestingly shows the use of technology to exploit the three cardinal axes of *frequency, time,* and *code*. First-generation systems were based on *Frequency Division Multiple Access (FDMA)*, where each user was assigned a single frequency or a pair of frequencies for full duplex communications. Second-generation systems expanded FDMA to *Time Division Multiple Access (TDMA)*, where each available frequency channel could transmit and receive data from several users by assigning each user data into individual time slots. Finally, third-generation systems introduced *Code Division Multiple Access (CDMA)*, which assigns orthogonal codes to subscribers: this enables them to share the entire available frequency band without interfering with each other.

In a sense, CDMA satisfied the need for large bandwidth and capacity, which enabled users to transmit and receive data with high speed and efficiency. However, increasing subscriber size and evolving internet applications keep driving the requirement for new developments in modulation and other communication technologies. While CDMA made the groundbreaking breakthrough in wideband communications, continuing generations such as *fourth-generation (4G)* and *fifth-generation (5G)* aim at optimizing system capacity and increasing data speed. The data itself consists of any information that we share in day-to-day life, like voice, text, email, and other forms of online communication.

6.1 Evolution of Capacity and Data Rate in Communication Systems

In Chapter 4, we defined *channel capacity* as the maximum possible bit rate of the communication system, for a given channel bandwidth. This is an important system factor related to practical performance, for example, the speed of your internet provider. Likewise, *system capacity* is also another equally important system factor that relates to the maximum number of users or *subscribers* possible in the system. All of us are subscribers, who pay regular charges to our internet service provider. The important question arises: *How does the service provider accommodate the thousands of users within*

the system? The following material will take us through the different generations of communication systems, and the techniques used to provide quality
service to the subscribers.

6.1.1 Frequency Division Multiple Access

First-generation communication systems utilized the simple principle of
assigning a frequency pair, or duplex channel, for each user demand. For
example, if B_T is the total system bandwidth available, and B is the duplex
channel bandwidth, then the total number of available channels, N, is

$$N_{FDMA} = \frac{B_T}{B} \tag{6.1}$$

If one duplex channel is assigned to each user, then the maximum number
of subscribers possible at the given time is also N. FDMA systems are simple,
but also have low bandwidth (~30 kHz) owing to channel division. In cellular systems, for example, *guard bands* are also inserted between adjacent
channels, to allow for nonideal filtering and other practical considerations.
In this case, Equation 6.1 becomes modified as given in Equation 6.2:

$$N_{FDMA} = \frac{B_T - 2B_G}{B} \tag{6.2}$$

This modification is to allow for guards bands, each of bandwidth B_G, at the
start and end of each subscriber channel.

Example

The U.S. FDMA system used a total bandwidth of 10 MHz, with 30 kHz
channels and guard bands of 10 kHz. What is the system capacity if each
user requires a duplex pair to transmit/receive?

Solution

Referring to Equation 6.2, the number of channels possible is

$$N_{FDMA} = \frac{B_T - 2B_G}{B}$$

$$= \frac{10 \text{ MHz} - 2(10 \text{ kHz})}{30 \text{ kHz}}$$

$$= \frac{10 \times 10^6 - 2 \times 10 \times 10^3}{30 \times 10^3} = 416 \text{ channels}$$

If each user requires a duplex pair of channels, then maximum number
of users possible is 416/2 = 208 users.

6.1.2 Time Division Multiple Access

As the number of wireless subscribers increased significantly, engineers had to develop a novel technology to accommodate this increased demand. TDMA essentially *allows two or more subscribers to use the same frequency channel*, but how is this possible without the users interfering with each other? The solution to this problem is digital technology or in hardware terms, a microprocessor, which can take digital data from many subscribers, and assign them into different *time slots* within a single *frame*. Figure 6.1 shows the structure of a TDMA system, which has M time slots in a single frame.

Each slot is assigned to a single user, but it is also possible to assign two slots to a single user: one slot to transmit and one to receive. The key aspect is that the frame repeats cyclically, so user information is sent in *bursts*, rather than continuously, on the same channel. Since there are L slots/frame, potentially a maximum of L users could be accommodated on a single frequency channel; hence, the system capacity gets multiplied by this factor as shown in Equation 6.3:

$$N_{TDMA} = L\left[\frac{B_T - 2B_G}{B}\right] \tag{6.3}$$

Since the TDMA transmitter breaks up the user data, and puts it back together at the receiver, additional information needs to be sent in each frame to accurately synchronize different slots and also frames. Hence, the frame content can be divided into *data bits* and *overhead bits*, which are used to transmit slot and frame synchronization information. The *frame efficiency*, or the percentage of information that is actually used to transmit data, is defined in Equation 6.4:

$$\eta = \frac{b_T - b_{OH}}{b_T} \times 100 \tag{6.4}$$

where b_T and b_{OH} are the total number of bits/frame and number of overhead bits/frame respectively.

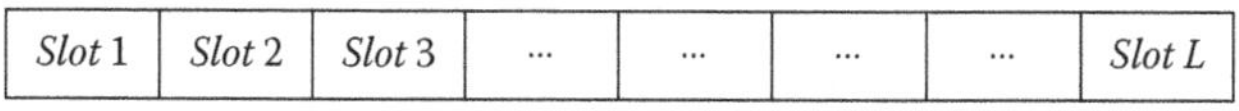

Slot 1	Slot 2	Slot 3	...	...	...	...	Slot L

FIGURE 6.1
Frame structure for GSM TDMA.

Example

The Global System for Mobile Communications; originally Groupe Spécial Mobile (GSM) uses a frame structure where each frame consists of eight time slots, and each time slot contains 156.25 bits, and data is transmitted at 270.833 kbps in the channel. Calculate the following:

(a) Time duration of a bit.
(b) Time duration of a slot.
(c) Time duration of a frame and frame efficiency, if 322 bits/frame are used as overhead.
(d) If the total system bandwidth is 10 MHz, with 25 kHz channels and 5 kHz guard bands, what is the system capacity if each user requires a duplex pair of time slots or frequency channels to transmit/receive?

Solution

(a) Time duration of a bit, $T_b = 1/\text{Bit rate}$
$$= 1/270.833 \text{ kbps} = 3.692 \text{ μs}.$$

(b) Time duration of a slot, $T_s = 156.25 \times T_b$
$$= 156.25 \times 3.692 \text{ μs} = 0.577 \text{ ms}.$$

(c) Time duration of a frame, $T_f = 8 \times T_s$
$$= 4.615 \text{ ms}.$$

From Equation 6.4, frame efficiency is calculated as

$$\eta = \frac{b_T - b_{OH}}{b_T} \times 100$$

$$= \frac{156.25 \times 8 - 322}{156.25 \times 8} \times 100$$

$$= 74.24\%$$

(d) Referring to Equation 6.3, the number of channels possible is

$$N_{TDMA} = L\left[\frac{B_T - 2B_G}{B}\right]$$

$$= 8\left[\frac{10\,\text{MHz} - 2(5\,\text{kHz})}{25\,\text{kHz}}\right]$$

$$= 8\left[\frac{10 \times 10^6 - 2 \times 5 \times 10^3}{25 \times 10^3}\right] \sim 3200 \text{ channels}$$

If each user requires a duplex pair of channels, then maximum number of users possible is 3200/2 = 1600 users.

6.1.3 Code Division Multiple Access

Unlike FDMA and TDMA systems, CDMA works on the principle of *sharing* the entire available bandwidth rather than *dividing* the bandwidth into channels. This is possible through spread spectrum technology, which was discussed in Chapter 5. Since there is no frequency division, ideally there is *no limit* on the number of users possible in a CDMA system. However, just as any room would start getting uncomfortable as more and more people turn up for the party, the number of users in a CDMA system is controlled by the *SNR* in the channel, and also the allowable *BER* of the system. The *BER* formulae given in Equations 5.3 and 5.6 for FHSS and DSSS systems, respectively, can be used to estimate the number of users for a given *SNR*.

The evolution of communication systems from 1G to 3G, as detailed in the earlier sections, gradually increased the system capacity (i.e., number of users), channel capacity (i.e., bit rate), and the bandwidth available for subscribers. However, wireless demand has grown exponentially and in the year 2000, a major threshold was reached: *Number of wireless users equaled the number of wired users.* In parallel with the growing number of wireless users, the need for faster data rates and higher bandwidth has fueled the innovation for even further generations of systems, which will be described in the following sections.

6.2 Fourth-Generation Systems

Long-Term Evolution (LTE) is the fourth-generation (4G) wireless communication standard for high-speed data of up to 200 Mbps, which is around 10 times faster than 3G systems. This standard was initiated by Nippon Telegraph and Telephone (NTT) DoCoMo in 2004, as part of the *3GPP* or *Third Generation Partnership Project*, which is an international collaboration between various telecommunication groups. The goal of LTE is to increase the capacity and speed of wireless data networks using the following approaches:

- New DSP and modulation techniques
- Simplification of the wireless network architecture

The LTE wireless interface is incompatible with 2G and 3G networks, so it must be operated on different wireless channels. LTE also uses a *different* radio technology, instead of CDMA that is used in 3G systems. This includes a new modulation technique called *Orthogonal Frequency Division Multiplexing (OFDM)*, and capacity improvement technique called *Multiple Input Multiple output (MIMO)*.

6.2.1 Multicarrier Approach to Modulation

To understand the need for new modulation technology in 4G LTE, let us take a look at the evolution of digital communication systems from second

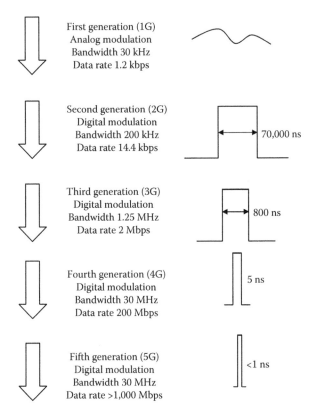

FIGURE 6.2
Evolution of data rate and pulse width with system generation.

generation to fourth generation, as shown in Figure 6.2. As can be seen, the bandwidth has increased significantly by 1000 times from around 30 kHz at 1G to 30 MHz Mbps at 4G! Correspondingly, the pulse width (=1/Data rate) of the basic information bit has gone down from ~70 μs (or 7000 ns) (for 2G systems) to ~5 ns (4G systems). The problem is that as pulses become narrower, they also become unstable and more susceptible to interference from noise and other signals appearing within the same wide bandwidth.

The solution was to move away from the single carrier system to a *multicarrier* system, where the high-speed data is divided between many *subcarriers*. Since each subcarrier has a lower information rate, the data pulses will be longer, making them more protected against noise and other interference. Additionally, if the subcarriers are *orthogonal*, it is still possible to recover the individual subcarriers' signals even if the subcarriers are overlapping. This principle is called as *OFDM*, and will be explained in detail in the next sections.

6.2.2 Principle of Orthogonal Frequency Division Multiplexing

Figure 6.2 shows a typical three-subcarrier system, with frequencies f_1, f_2, and f_3. Each subcarrier's frequency spectrum is represented by a sinc function, one of whose properties is to *peak at its center frequency and go to zero at all integer multiples of this frequency*. Figure 6.3 shows the multiples of the three subcarrier frequencies, and it is important to note that they are packed closely together, but with a specific pattern. Using the latter pattern, the OFDM receiver can effectively demodulate each subcarrier because, at the peaks of each of these sinc functions, the contributions from other subcarrier sinc functions are zero.

6.2.3 OFDM Transmission and Reception

This modulation technique divides a channel with large bandwidth (5–10 MHz) into smaller *subchannels* each 15 kHz wide. Each is modulated with part of the data. The fast data is divided into slower streams that modulate the subcarriers with one of several modulation schemes like QPSK or 16QAM. Figure 6.4a shows a typical OFDM transmitter. The input signal, $s(n)$, is a serial stream of binary digits. This composite signal is first demultiplexed into N parallel streams, and each one is mapped to a symbol stream using some modulation format (QAM, PSK, etc.). An inverse FFT is computed

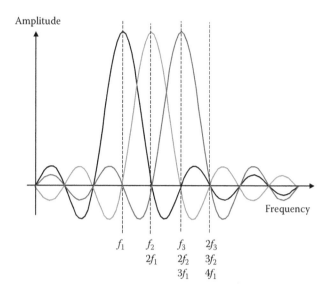

FIGURE 6.3
Three-subcarrier system in OFDM.

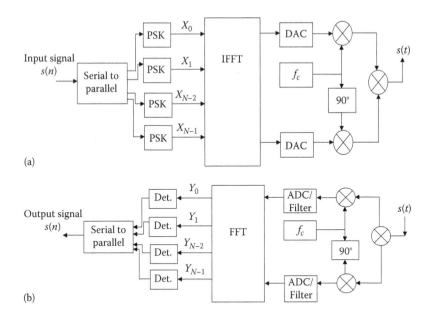

(a)

(b)

FIGURE 6.4

(a) Block diagram of OFDM transmitter. (b) Block diagram of OFDM receiver.

on each set of symbols, giving a set of complex time-domain samples, as described in Equation 6.5:

$$y(t) = \frac{1}{\sqrt{N}} \sum_{k=0}^{N-1} X_k e^{j2\pi f_k t}, \quad 0 < t < NT \tag{6.5}$$

where

f_k is the frequency of the kth subcarrier

T is the base symbol duration

NT is the length of the OFDM symbol

The subcarrier frequencies f_k are equally spaced, with values given by Equation 6.6:

$$f_k = \frac{k}{NT} \tag{6.6}$$

These samples are then quadrature-mixed to passband in the standard way. The real and imaginary components are first converted to the analog domain using a digital-to-analog converter (DAC); the analog signals are then used to modulate cosine and sine waves at the carrier frequency, f_c, respectively. These signals are then summed to give the transmission signal, $s(t)$. Figure 6.4b shows a typical OFDM receiver.

The receiver picks up the signal $r(t)$, which is then quadrature-mixed down to baseband using cosine and sine waves at the carrier frequency. This also creates signals centered on $2f_c$, so low-pass filters are used to reject these. The baseband signals are then sampled and digitized using analog-to-digital converter (ADC), and a forward fast Fourier transform (FFT) is used to convert back to the frequency domain. This returns N parallel streams, each of which is converted to a binary stream using an appropriate symbol detector. These streams are then re-combined into a serial stream, $\hat{s}(n)$, which is an estimate of the original binary stream at the transmitter.

6.2.4 Advantages and Disadvantages of OFDM

The main advantages of OFDM are as follows:

- *High spectral efficiency*, as compared to other modulation techniques, such as double sideband modulation (DSB) in 2G and spread spectrum techniques in 3G systems
- *Efficient and fast implementation*, since it uses the FFT
- *Resistant to intersymbol interference (ISI)* and fading due to multipath propagation

Disadvantages include the following:

- Since the subcarriers have to be precisely placed to ensure orthogonality, *frequency synchronization* problems are possible.
- Sensitive to *Doppler shift*.
- *High peak-to-average power ratio (PAPR)*, which requires poorly efficient linear transmitter circuitry.

6.3 Multiple-Input Multiple-Output Technology

MIMO is a radio communications technology that is currently used predominantly in high bandwidth systems to provide increased capacity, or number of users, and also channel reliability. MIMO technology has been adopted in multiple wireless systems, including Wi-Fi, WiMAX, LTE, and is proposed for future standards (such as LTE-Advanced and IMT-Advanced). The latter improvements are obtained by the use of *multiple antennas* at the transmitter and receiver to increase the number of signal paths to carry the data.

6.3.1 Principle of MIMO Systems

To understand the MIMO advantage, let us compare the single-antenna system and the four-antenna MIMO system, shown in Figure 6.5a and b, respectively. Let us say that the transmitter has 20 frequency channels:

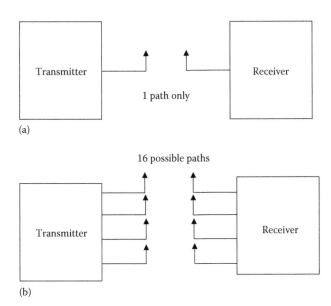

FIGURE 6.5
(a) Single-input, single-output system. (b) Four-input, four-output system.

in the single-input, single-output system, since there is only *one transmission path*, this would mean that only 20 users could transmit simultaneously. However, in the four-input, four-output system, since there are *16 different paths*, users can use both frequency and space to increase capacity. Hence, a maximum of $16 \times 20 = 320$ users can potentially use the system simultaneously. Of course, this increased capacity system would require more hardware, in the form of additional antennas and processors at the receiver to separate out all user data.

6.3.2 Analysis of Input–Output Systems

Various types on input–output systems are listed and briefly described in Table 6.1.

TABLE 6.1

Types of Input–Output Systems

SISO	Single-input, single-output: transmitter and receiver of the radio system have only one antenna.
SIMO	Single-input, multiple-output: receiver has multiple antennas, while the transmitter has one antenna.
MISO	Multiple-input, single-output: transmitter has multiple antennas, while the receiver has one antenna.
MIMO	Multiple-input, multiple-output: both the transmitter and receiver have multiple antennas.

Consider a MIMO transmitter with M *transmit* antennas, and a receiver with N *receive* antennas. The channel can be represented by the $N \times M$ matrix H of *channel gains* h_{ij} representing the gain from transmit antenna j to receive antenna i. The $N \times 1$ received signal $\mathbf{y}$ is given by Equation 6.7:

$$\mathbf{y} = H\mathbf{x} + \mathbf{n} \tag{6.7}$$

where
   $\mathbf{x}$ is the $M \times 1$ transmitted vector
   $\mathbf{n}$ is the $N \times 1$ additive white Gaussian noise vector

The channel capacity (or maximum bit rate in bps) of a MIMO system is given by Equation 6.8:

$$\text{Channel Capacity, } C = B \log_2 \left(\det \left[I + \frac{SNR}{M} HH^T \right] \right) \tag{6.8}$$

where
   B is the channel bandwidth
   I is the $N \times N$ identity matrix
   SNR is the signal-to-noise ratio (magnitude)
   M is the number of transmitter antennas
   N is the number of receiver antennas
   H is the $N \times M$ channel gain matrix
   "det" represents the determinant function

Note: For SISO, SIMO, and MISO systems, as defined in Figure 6.5, the channel capacity is defined by Shannon's maximum capacity theorem $C = B \log_2(1 + SNR)$, which was discussed in Chapter 3.

Example

(a) Calculate the channel capacity of a SISO system, operating in a wireless channel with bandwidth 1.2 MHz and SNR of 15 dB.
(b) Repeat part (a) for a MIMO system, with same channel bandwidth and SNR, and the following properties:

Number of transmitter antennas, $M = 3$

Number of receiver antennas, $N = 3$

$$\text{Channel gain matrix } H = \begin{bmatrix} 0.33 & 0.50 & 0.80 \\ 0.50 & 0.44 & 0.60 \\ 0.8 & 0.60 & 0.66 \end{bmatrix}$$

Solution

(a) Using Shannon's formula for SISO system:

$$C = B \log_2(1 + SNR)$$

Given $B = 1.2$ MHz, $SNR = 15$ dB $=> 10^{15/10} = 31.62$, we obtain:

$$C = 10^6 \log_2(1 + 31.62)$$

$$C = 5.03 \text{ Mbps}$$

(b) Using Equation 6.4 for MIMO system:

$$C = B \log_2\left(\det\left[I + \frac{SNR}{M} HH^T\right]\right)$$

$$I = \begin{bmatrix} 1 & 0 & 0 \\ 0 & 1 & 0 \\ 0 & 0 & 1 \end{bmatrix}$$

Then,

$$I + \frac{SNR}{M} HH^T$$

$$= \begin{bmatrix} 1 & 0 & 0 \\ 0 & 1 & 0 \\ 0 & 0 & 1 \end{bmatrix} + \frac{31.62}{3} \begin{bmatrix} 0.33 & 0.50 & 0.80 \\ 0.50 & 0.44 & 0.60 \\ 0.8 & 0.60 & 0.66 \end{bmatrix} \begin{bmatrix} 0.33 & 0.50 & 0.80 \\ 0.50 & 0.44 & 0.60 \\ 0.8 & 0.60 & 0.66 \end{bmatrix}$$

$$= \begin{bmatrix} 11.5284 & 9.1171 & 11.5097 \\ 9.1171 & 9.4699 & 11.1724 \\ 11.5097 & 11.1724 & 16.1312 \end{bmatrix}$$

and $\det\left(I + \frac{SNR}{M} HH^T\right) = 71.4835$.

Hence, the channel capacity of the MIMO system is given by

$$C = 10^6 \log_2(71.4835) = 6.16 \text{ Mbps}$$

Comparing, we see that the channel capacity of the MIMO system is almost 23% higher than that of the SISO system.

6.4 Fifth-Generation Communication Systems

The vision of fifth generation (5G) broadly encompasses the following network goals:

- *Highly efficient mobile network* that delivers a better performing network for lower investment cost. It would satisfy the mobile operators' current requirement of data rate cost falling at roughly the same rate as increasing demand for data volume.
- *Very fast mobile network* comprising the next generation of small cells densely clustered together to give neighboring coverage, over at least highly dense urban areas. It would require access to spectrum under 4 GHz perhaps via the world's first global implementation of *Dynamic Spectrum Access (DSA)*. DSA is a new spectrum-sharing model that allows secondary users to access the abundant spectrum holes or white spaces in the licensed spectrum bands.
- *Coordinated fiber-wireless network* that uses the millimeter wave bands (20–60 GHz) so as to allow very wide bandwidth radio channels to support data access speeds of up to 10 Gbps.
- *Short wireless links* like Wi-Fi, with local fiber-optic terminals rather than long range cellular service.

Problem Solving

P6.1

(a) Calculate the channel capacity of a SISO system, operating in a wireless channel with bandwidth 1.5 MHz and *SNR* of 12 dB.

(b) Repeat part (a) for a MIMO system, with same channel bandwidth and SNR, and the following properties:

Number of transmitter antennas, $M = 4$

Number of receiver antennas, $N = 5$

$$\text{Channel gain matrix } H = \begin{bmatrix} 0.33 & 0.50 & 0.80 & 0.50 \\ 0.50 & 0.44 & 0.60 & 0.44 \\ 0.80 & 0.60 & 0.66 & 0.60 \\ 0.14 & 0.25 & 0.75 & 0.44 \\ 0.37 & 0.55 & 0.42 & 0.60 \end{bmatrix}$$

P6.2
Next Generation Mobile Networks Alliance (NGNM) defines the 5G network requirements as given in the following. Please fill in an estimate number in the blanks:

(a) Data rates of almost 100 Mb/s should be supported for approximately _____ users.

TABLE 6.2

OFDM Components and Equivalent Simulink® Blocks

OFDM System Component	Corresponding Simulink Block
Serial to parallel/parallel to serial	Pulse generator at 1 MHz bit rate
PSK/Detector	PSK modulator/PSK demodulator
ADC/DAC	Analog to digital/digital to analog convertor
Carrier f_c	Sine wave at 900 MHz
90° phase shifter	Phase shift or delay

Notes: Use the following basic circuits to replicate PSK and detector if Simulink blocksets are not available.
For PSK modulator/demodulator, refer to Problem C4.4.
Scale down frequencies by a factor of 10 to reduce simulation time.

(b) 1 Gbps to be offered, simultaneously to approximately _____ on the same office floor.
(c) Up to _____ simultaneous connections to be supported for massive sensor deployments.
(d) Spectral _____ should be significantly enhanced compared to 4G.
(e) Signaling _____ and _____ should be improved.

Computer Laboratory

C6.1 Develop a Simulink® model to implement the OFDM transmitter in Figure 6.3a, using Table 6.2, showing blocks from the Simulink library to represent system components.

C6.2 Develop a Simulink model to implement the OFDM receiver in Figure 6.3b, using the same blocks as the transmitter model in Problem C6.1.

7

Long-Range and Short-Range Communication Networks

Typically, communication systems, ranging from 1G to 4G, are *long-range* links, covering hundreds or even thousands of kilometers. These links are established by different channel media such as coaxial, microwave, cable, satellite, and more recently, cellular networks. In parallel, *short-range* communication systems were also developed to provide connectivity within a limited distance. Short-range communication systems can be classified broadly into the following categories:

- *Wireless local area networks (WLANs)*, such as *Wi-Fi*
- *Personal area networks (PANs)* such as *Bluetooth* and *Zigbee*
- *Ultra-wideband (UWB)* carrierless networks

Figure 7.1 gives an overall picture of short-range networks and applications.

7.1 Wireless Local Area Networks (WLANs)

A wireless local area network (WLAN) is a wireless network that links many devices within a limited area such as a home, a school, a university or an office building. This gives users the ability to move around within the local coverage area and still be connected to the network. Modern WLANs use either spread spectrum or orthogonal frequency division multiplexing (OFDM) technology, under the IEEE 802.11 standard, and are marked under the *Wi-Fi (Wireless Fidelity)* brand name. Norman Abramson, a professor at the University of Hawaii, developed the world's first wireless computer communication network, *ALOHAnet* in 1971, using seven computers deployed over four islands to communicate with the central computer on Oahu Island.

WLAN systems are similar to cellular systems, where the base stations are replaced by *access points (APs)*, and cell phone users are replaced by clients such as laptops and desktops, which are equipped with *wireless network interfaces*. WLANs use OFDM technology in the 2.5 and 5 GHz frequency bands to connect to the local APs, which in turn are connected to the large area networks, which are formally called as Wide Area Networks (WLANs) maintained by satellite, microwave and fiber optic and cellular links.

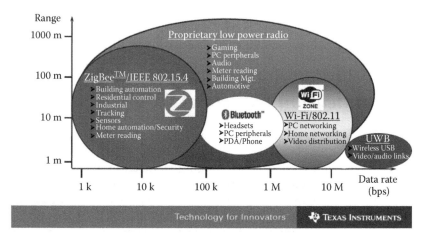

FIGURE 7.1
Overview of short-range wireless networks. (Courtesy of Texas Instruments, Dallas, TX.)

7.1.1 Types of WLAN Specifications

The 802.11 and 802.11 ext refer to the range of specifications, developed by the IEEE for WLAN technology. The specific standard definitions are as follows:

- 802.11 applies to WLANs and provides up to 2 Mbps transmission in the 2.4 GHz band using either Frequency Hopping Spread Spectrum (FHSS) or Direct Sequence Spread Spectrum (DSSS).

- 802.11a is an extension to 802.11 that applies to WLANs and provides up to 54 Mbps in the 5 GHz band. 802.11a uses OFDM encoding scheme rather than FHSS or DSSS.

- 802.11b (*also referred* to as *802.11 high rate* or *Wi-Fi*) is an extension to 802.11 that applies to wireless LANS and provides 11 Mbps transmission in the 2.4 GHz band. 802.11b uses only DSSS and was added as a 1999 ratification to the original 802.11 standard, allowing wireless functionality comparable to Ethernet.

- 802.11e is a wireless draft standard that defines the *quality of service (QoS)* support for LANs, and is an enhancement to the 802.11a and 802.11b WLAN specifications, while maintaining full backward compatibility with these standards.

- 802.11g applies to WLANs and is used for transmission over short distances at up to 54 Mbps in the 2.4 GHz bands.

- 802.11n builds upon previous 802.11 standards by adding *Multiple-Input Multiple-Output (MIMO)*. The additional transmitter and receiver antennas allow for increased data throughput (~100 Mbit/s) through spatial multiplexing, and increased range by exploiting the spatial diversity through coding schemes like *Alamouti* coding.

- 802.11ac builds upon the 802.11n standard, to deliver data rates of 433 Mbps per spatial stream, or 1.3 Gbps in a three-antenna (three stream) design. The 802.11ac specification operates only in the 5 GHz frequency range.

- 802.11ad is a wireless specification with an operating frequency in the 60 GHz frequency band and offers maximum data rate of up to 7 Gbps.

- 802.11r is also called *Fast Basic Service Set (BSS)* Transition, supports handoff between APs to enable *Voice Over Internet Protocol (VOIP)* roaming on a Wi-Fi network.

- 802.1X is an IEEE standard for port-based network access control that allows network administrators restricted use of IEEE 802 LAN service APs, to secure communication between authenticated and authorized devices.

7.1.2 Wi-Fi Networks

Wi-Fi Alliance was founded in 1999, and has since developed an interoperability certification program to ensure that Wi-Fi devices will work together efficiently without any compatibility problems. It defines Wi-Fi as any WLAN product based on the IEEE 802.11 standard. The "Wi-Fi-certified" trademark can only be used by Wi-Fi products that successfully complete Wi-Fi Alliance interoperability certification testing.

Many devices can use Wi-Fi; for example, personal computers, video-game consoles, smartphones, and digital cameras. These can connect to a network resource such as the Internet, via wireless APs. Such an AP (or hotspot) has a range of about 20 m indoors and a greater range outdoors. Hotspot coverage can comprise an area as small as a single room with walls that block radio waves, or as large as many square kilometers achieved by using multiple overlapping APs.

Wi-Fi usage has spread significantly, especially in university campuses, shopping malls, airports and most public places. Citywide service started in the United States in 2005, with Sunnyvale, California being the first to offer free Wi-Fi. However, large-scale expansion of free citywide Wi-Fi is still limited, considering the cost involved in setup and maintenance and also competition from local communication industry, which provides superior service at a reasonable cost.

7.1.2.1 Wi-Fi Hotspots and Network Types

A *hotspot* is basically an area with an accessible wireless network. The term is most often used to refer to wireless networks in public areas like

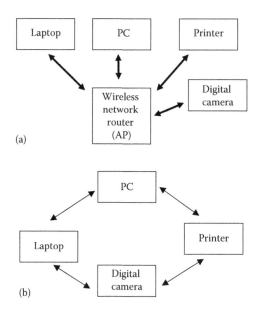

FIGURE 7.2
(a) Infrastructure network for Wi-Fi. (b) *Ad hoc* network for Wi-Fi.

airports and coffee shops; these networks may be free for use or may carry a fee. There are generally two types of Wi-Fi wireless networks: An *infrastructure network*, as shown in Figure 7.2a, is a wireless network configuration that is commonly used in home networks and hotspots. Data transferred between wireless devices on the network pass through a central *AP*, such as a *wireless network router*. An *ad hoc* network is a wireless network where data is transmitted directly between wireless devices on the network, without passing through an AP. Such a network is shown in Figure 7.2b.

7.1.2.2 Advantages of Wi-Fi

- Wi-Fi is a very *economical* solution to provide WLANs, especially in small outdoor areas.
- Almost all laptops and PDAs come equipped with wireless network adapters, thus providing seamless connectivity between people.
- Products designated as "Wi-Fi certified" are backward *compatible* and any standard Wi-Fi device will work anywhere in the world.
- Finally, *Wi-Fi Protected Access Encryption* (WPA2) provides good security in networks.

7.2 Personal Area Networks (PANs)

PANs are computer networks that link devices such as computers, telephones, and personal digital assistants. PANs can be used for communication among personal devices that are within a few meters of each other, or to connect to a higher level network and the Internet (an uplink). Examples of *Wireless Personal Area Networks (WPANs)* include *Bluetooth* and *Zigbee*, and are defined by the IEEE 802.15 standard.

7.2.1 Bluetooth

Bluetooth is a wireless technology standard that was invented by the telecom company, Ericsson, in 1994. It is based on the *IEEE 802.15.1* standard, and is managed by the *Bluetooth Special Interest Group (SIG)*. The Bluetooth SIG oversees development of the specification, manages the qualification program, and protects the trademarks. A manufacturer must make a device meet Bluetooth SIG standards to market it as a Bluetooth device.

7.2.1.1 Potential of Bluetooth Technology

Bluetooth functions with FHSS wireless technology that uses short-range radio waves in the ISM frequency band of 2.4–2.485 GHz to connect devices such as keyboards, pointing devices, audio headsets, printers, personal digital assistants (PDAs), and cell phones. A Bluetooth PAN is also called a *piconet* (very small network) that typically has a range of 10 m. It is usually composed of up to eight active devices in a *master–slave* relationship; the first Bluetooth device in the piconet is the master, and all other devices are slaves that communicate with the master.

Hence, Bluetooth has the potential to replace all cables linking devices such as computers, printers, and keyboards. The Bluetooth name originates from the tenth century king, *Harald Bluetooth*, who united the Scandinavian countries, just as our Bluetooth network connects all our wireless devices. Figure 7.3a and b depicts the system blocks that go into a Bluetooth transmitter and receiver, respectively. In the *transmitter*, the input data at a rate of 1 MHz is sent through the *Gaussian frequency shift keying (GFSK)* filter to smoothen out the pulses. The filter has a 3 dB cutoff frequency of 500 kHz, and adds the term Gaussian to the normal FSK modulation process. The PN hop sequence generator outputs a 16 kHz pulse that varies periodically over 79 amplitude levels in the range of (–1 V, 1 V) with the purpose of hopping the voltage-controlled oscillator (VCO) frequency in the range 2402–2462 MHz. The Bluetooth signal is passed through a *wideband filter* before being transmitted via the antenna.

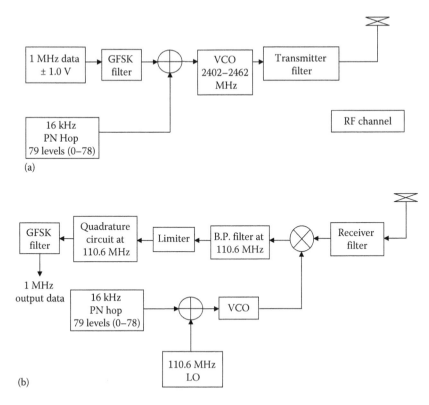

FIGURE 7.3
Bluetooth (a) transmitter and (b) receiver.

In the receiver, after passing through the wideband receiver filter, the high-frequency signal is downconverted to an IF range, centered at 110.6 MHz. The downconversion is achieved by mixing the received signal with a frequency hopped VCO signal in the range of 2512.6–2590.6 MHz, which gives the desired IF frequency of 110.6 MHz.

7.2.1.2 Applications of Bluetooth

- Wireless control of and communication between a mobile phone and a remote headset.
- Wireless control of and communication between a mobile phone and a Bluetooth compatible car stereo system.
- Wireless communication with PC input and output devices; the most common are the mouse, keyboard, and printer.
- Replacement of previous wired RS-232 serial communications in test equipment, GPS receivers, medical equipment, bar code scanners, and traffic control devices.

- Wireless bridge between two industrial Ethernet networks.
- Game consoles such as Nintendo's Wii and Sony's PlayStation, use Bluetooth for their respective wireless controllers.
- Dial-up Internet access on personal computers or PDAs using a data-capable mobile phone as a wireless modem.
- Short-range transmission of health sensor data from medical devices to mobile phone, or dedicated telehealth devices.
- Real-time location systems (RTLS) are used to track and identify the location of objects in real-time using *nodes* or *tags* attached to, or embedded in the objects tracked, and *readers* that receive and process the wireless signals from these tags to determine their locations.
- Personal security application on mobile phones for prevention of theft or loss of items.
- In some countries like Canada, data collected from travelers' Bluetooth devices is used to predict travel times and road congestion for motorists.
- Wireless transmission of audio, as a more reliable alternative to FM transmitters.

7.2.2 ZigBee

ZigBee is the right term. is an open, global wireless system based on the *IEEE 802.15.4* standard, and is a PAN. The ZigBee name originates refers to the *waggle dance* of honeybees after their return to the beehive. Since it is a low-power network, transmission distances are limited to 10–100 m; however, ZigBee devices can transmit data over long distances by passing data at 250 kbps through a *mesh network* of repeater devices. ZigBee operates in the *Industrial, Scientific and Medical (ISM)* radio bands: 2.4 GHz in most jurisdictions worldwide, 784 MHz in China, 868 MHz in Europe, and 915 MHz in the United States and Australia.

7.2.2.1 Potential of ZigBee Technology

ZigBee is best suited for intermittent data transmissions from a sensor or input device, such as temperature or rain sensors used in agricultural applications. Ultimately, ZigBee is a standard to support the *Internet of Things* by enabling simple and smart objects to work together, improving comfort and efficiency in everyday life. The ZigBee Alliance is a nonprofit association of approximately 400 members driving development of innovative, reliable, and easy-to-use ZigBee standards. The Alliance promotes worldwide adoption of ZigBee as the leading wirelessly networked, sensing, and control standard for use in consumer, commercial, and industrial areas.

ZigBee technology has *lower installation and hardware cost*, and *lower power consumption* than other WPANs, such as Bluetooth. ZigBee systems provide *Low Rate Wireless Personal Area Networks (LR-WPANs)* in residential and industrial environments. *Advantages and disadvantages* of ZigBee systems include the following:

- Complexity of heavy protocol stacks
- Low throughput: 250 kbps
- Low module cost
- Low installation cost
- Low power consumption

7.2.2.2 Applications of ZigBee

- Public safety, such as *sensing and location determination* at disaster sites
- Automotive sensing, such as *tire pressure monitoring*
- Monitoring agriculture, such as *sensing of soil moisture, pesticide, rain content*
- Home monitoring and control, such as *heating, air conditioning (HVAC), and security*
- Consumer electronics such as *remote controls for radio, TVs, DVDs*
- Health monitoring such as sensors and monitors for *vital functions*

One important application system is the *ZigBee Smart Energy*, which utilizes an *IP-based protocol* to monitor and control the delivery and use of energy and water. It includes services for *plug-in electric vehicle (PEV)* charging, installation, prepay services, and user information and messaging, through a connection of wired and wireless networks.

7.3 Ultra-Wideband Systems

The concept of the *carrier* signal has always been the backbone of electronic communications, from first-generation analog to fourth-generation LTE systems. Typically, the carrier is always a *high-frequency* sinusoid signal, and the message is the *low-frequency* signal. Now, what if the message itself was very high frequency? This is basically the principle behind *Ultra-wideband* (UWB) systems, pioneered by Robert A. Scholtz and others, where a high-frequency *pulse* conveys the information, without the need for an additional sinusoidal carrier. Hence, while typical 3G or 4G communications is carrier controlled and wideband (~100 MHz), a UWB system is *carrierless* and wideband (~500 MHz).

7.3.1 Frequency Bandwidth of UWB Systems

UWB systems transmit information using very narrow pulses spread over a large bandwidth of 500 MHz or more. This extremely high bandwidth ensures that UWB can be used only for small range networks, to prevent interference with cellular and other *Wide Area Networks (WANs)*. Pulse-based UWB radars and imaging systems tend to use low repetition rates (typically in the range of 1–100 megapulses/s). On the other hand, communications systems favor high repetition rates (typically in the range of 1–2 gigapulses/s), thus enabling short-range gigabit-per-second communications systems. Each pulse in a pulse-based UWB system occupies the entire UWB bandwidth, thus enjoying the benefits of higher immunity to *multipath fading*, unlike *carrier-based* systems that are susceptible to deep fading and intersymbol interference.

7.3.2 UWB Transmission and Reception

Figure 7.4 shows the block diagram of the typical UWB communication system. As can be seen, UWB systems are remarkably simple in architecture. The *transmitter* consists of just the high-frequency *pulse generator*, which is modulated by the input data, and *transmitting antenna*. Typical pulse modulation techniques include *pulse amplitude modulation (PAM)*, *pulse width modulation (PWM)*, and *pulse position modulation (PPM)*, as shown in Figure 7.5. In PPM, the *pulse amplitude* varies according to the data amplitude, $g(t)$; in PWM, the *pulse width* varies according to the data amplitude, and finally in PPM, the position of the pulse, defined with respect to the uniform sample points, varies according to the data amplitude.

The UWB *receiver* consists of *receiving antenna*, the *low noise amplifier (LNA)*, the *data extract* block, which extracts the data from the pulse modulated signal, and finally, the *correlator*, which is usually a matched filter that the message by making the high or low (1 or 0) data decision. Additionally,

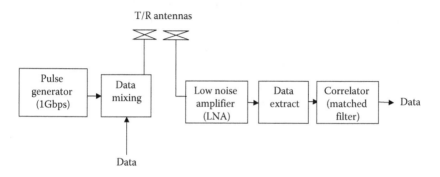

FIGURE 7.4
Block diagram of UWB transmitter/receiver.

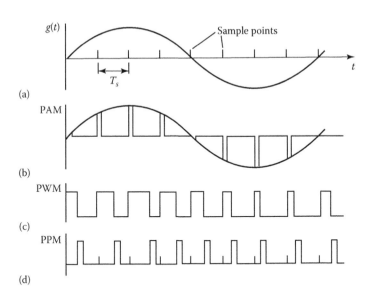

FIGURE 7.5
Pulse modulation schemes.

unlike conventional carrier-based modulation systems, there is no requirement for power amps, transmit filters, VCO, mixer, or local oscillators.
This simple hardware structure directly relates to *low-cost* and *low-power*
consumption.

7.3.3 UWB Pulse Generation

The pulse is the key component in UWB transmission and reception. To
clearly define and put limits on the pulse design, the FCC created a mask,
as shown in Figure 7.6, which plots the *power spectral density* (dBm/MHz)
as a function of frequency. The mask specifies the time and frequency
boundaries for candidate pulse designs. The FCC mask specifies the average bandwidth of UWB systems to be ~7.5 GHz (7500 MHz), as compared
to a standard 3G application such as Bluetooth, which has a much lower
bandwidth of ~200 MHz. The fractional bandwidth, BW_F, of a communication system can be defined as

$$BW_F = 2\frac{f_H - f_L}{f_H + f_L} = \frac{f_H - f_L}{f_C} \tag{7.1}$$

where f_H, f_L, and f_C are defined as the upper, lower, and center frequencies,
respectively, of the system. While conventional 3G systems rarely exceed a
fractional bandwidth of 50%, UWB systems reach almost 99%. Based on the

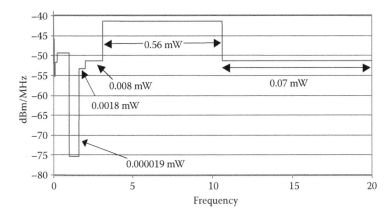

FIGURE 7.6
FCC specification mask for UWB.

FCC requirements, several optimal pulse candidates have been developed, as given below:

- Gaussian impulse
- Monocycle
- Doublet

These UWB pulse candidates are shown in Figure 7.7, in both time and frequency domains.

7.3.4 Potential Advantages and Disadvantages of UWB

Advantages of UWB include the following:

- Low cost, low power: simple implementation
  - Carrierless, direct baseband signal
  - Low duty cycle operation
- Potential for high capacity: high throughput
  - Large effective processing gain
  - Share the spectrum with many users
- Good propagation quantities
  - Multipath resistant, cm location
  - High penetration (high BW, low frequency)

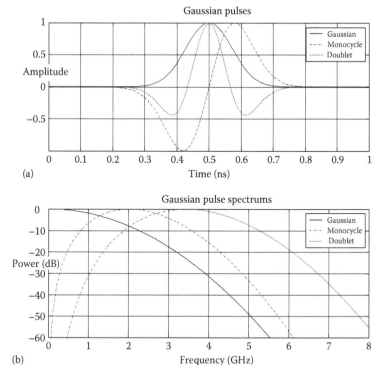

FIGURE 7.7
(a) Time and (b) frequency plots of UWB pulses.

Some potential issues include

- Regulatory
  - Noise aggregation issues
  - Wireless Internet connectivity issues
- Performance and implementation
  - Pulse synchronization
  - Susceptibility to interference
- Short range (a few meters to a few km)
  - Low power direct pulse operation
  - Low antenna transmit efficiency

7.3.5 Applications of UWB

- Imaging systems (medical, surveillance, ground penetrating radar), which may operate either below 960 MHz or between 1.99 and 10 GHz

- Vehicular radar systems (above 24.075 GHz)
- Communications and measurements systems restricted to indoor networks or hand-held devices

Example

An UWB system operates in the frequency range of 3.1–10.6 GHz. Determine the following:

(a) Fractional bandwidth of the system.
(b) Channel capacity of the system operating with SNR of 20 dB.

Solution

(a) From Equation 7.1, fraction bandwidth is calculated as

$$BW_F = 2\frac{f_H - f_L}{f_H + f_L}$$

$$= 2\frac{10.6 - 3.1}{10.6 + 3.1}$$

$$= 1.1 \text{ or } 110\%!$$

(b) Channel capacity is given by Shannon's theorem (Equation 4.11)

$$C = B \log_2(1 + SNR)$$

Using $B = 10.6 - 3.1 = 7.5$ GHz, and $SNR = 20$ dB => 100, we obtain

$$C = 7.5 \times 10^9 \log_2(1 + 100)$$

$$= 499.37 \text{ Gbps}!$$

Both fractional bandwidth and channel capacity are extremely high; however, range of the network is limited.

7.4 Path Loss Calculations in Long-Range and Short-Range Networks

In Chapter 1, we defined *path loss* in a communication system as the ratio of the transmitted power to the received power. It is important to estimate path loss in wired or wireless networks for the following reasons:

- Transmitter parameters like power, frequency, and bandwidth can be designed accurately to ensure that subscribers receive sufficient signal power to operate their communication devices.

- Path loss can vary significantly depending on the channel; for example, a signal traveling over free space or through a fiber-optic cable can be very predictable, as compared to a signal traveling within a building, or even through a city with many buildings.

7.4.1 Line of Sight (LOS) Model

The Line of Sight (LOS) or free-space-propagation model predicts the received signal's strength when the transmitter and the receiver have a *clear, unobstructed line-of-sight path* between them, as shown in Figure 7.8. The free-space power that the receiver's antenna receives a distance d from the transmitter is given by the Friis formula, in the following equation:

$$P_r = \frac{P_t G_t G_r \lambda^2}{(4\pi d)^2} \tag{7.2}$$

where
    P_t is the transmitted power in Watts
    P_r is the received power in Watts
    G_t is the gain of the transmitting antenna
    G_r is the gain of the receiving antenna
    λ is the wavelength in meters
    d is the transmitter–receiver distance in meters

Hence, from our earlier definition, the path loss (*PL*) for LOS propagation can be written as follows in Equation 7.3 below:

$$PL = \frac{P_t}{P_r} = \frac{(4\pi d)^2}{G_t G_r \lambda^2} \tag{7.3}$$

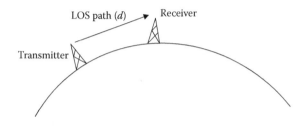

FIGURE 7.8
LOS propagation model.

and the logarithmic model is given in Equation 7.4 below:

$$PL(\text{dB}) = P_t(\text{dBm}) - P_r(\text{dBm})$$

$$= 20\log(4\pi d) - G_t(\text{dB}) - G_r(\text{dB}) - 20\log(\lambda) \tag{7.4}$$

In a cellular environment, for example, the *maximum range*, d_{max}, of a LOS system can be predicted based on the *Minimum Usable Level (MUL)* of the receiver. *MUL* defines the minimum signal power that is required to operate a wireless device like a cell phone.

Example

Assuming a transmitting frequency of 900 MHz, a transmit power of 10 W, and *unity gain* transmitting and receiving antennas:

(a) Determine the received power at 1 km distance in an outdoor LOS environment.
(b) What would be the received power and path loss (dB) at 2 km distance?
(c) What is the maximum range possible if the receiver *MUL* is −60 dBm?

Solution

(a) From Equation 7.2, received power is given by the Friis equation:

$$P_r = \frac{P_t G_t G_r \lambda^2}{\left(4\pi d\right)^2}$$

$$P_t = 10\,\text{W}$$

$$G_t = 1$$

$$G_r = 1$$

$$\lambda = \frac{c}{f}$$

where
c is the velocity of light in m/s
f is the frequency in Hz

Putting $c = 3 \times 10^8\,\text{m/s}$, and $f = 900$ MHz, and substituting all the values in the Friis equation, we obtain

$$P_r = 7.036 \times 10^{-9}\,\text{W}$$

(b) To obtain the power at 2 km, we can use the power at 1 km as follows (an extension of the Friis formula):

$$\frac{P_r(2\,\text{km})}{P_r(1\,\text{km})} = \frac{(1\,\text{km})^2}{(2\,\text{km})^2}$$

$$\text{or,}\quad P_r(2\,\text{km}) = \frac{(1\,\text{km})^2}{(2\,\text{km})^2} P_r(1\,\text{km})$$

$$= \frac{P_r(1\,\text{km})}{4} = \frac{7.036 \times 10^{-9}}{4}$$

$$= 1.759 \times 10^{-9}\,\text{W}$$

From Equation 7.4, the path loss is given by

$$PL(\text{dB}) = 20\log(4\pi d) - G_t(\text{dB}) - G_r(\text{dB}) - 20\log(\lambda)$$

$$= 20\log(4\pi \times 2000) - 0 - 0 - 20\log\left(\frac{3 \times 10^8}{900 \times 10^6}\right)$$

$$= 97.55\,\text{dB}$$

(c) $MUL = -60$ dBm

$\Rightarrow 10^{-60/10} = 10^{-6}$ mW or 10^{-9} W.

Using Equation 7.2, we obtain

$$P_r = \frac{P_t G_t G_r \lambda^2}{\left(4\pi d\right)^2}$$

$$10^{-9} = \frac{10 \times 1 \times 1 \times \left((3 \times 10^8)/(900 \times 10^6)\right)^2}{\left(4\pi d\right)^2}$$

and solving, the maximum range $d = d_{\max} = 2652.58$ m.

7.4.2 Practical Channel Models

As seen in the earlier section, LOS propagation is very stable, and also very predictable process. However, in many practical channels like cellular links within a city, LOS may not be the only path between the transmitter and receiver; in some cases, LOS may not even exist due to the presence of high buildings. Other propagation mechanisms include the following:

- *Reflection* when the signal bounces off from buildings and ground, which are much larger than the wavelength of the signal

- *Diffraction* or bending of the signal around sharp corners or rounded surfaces, like the curvature of earth
- *Scattering* of the signal by objects such as foliage, snow, and rain, which are smaller than the wavelength of the signal

Physical models exist for the three propagation methods defined earlier; however, in a city, for example, it is not practically possible to characterize all the reflection, diffraction, and scattering from the hundreds of buildings, trees, and other objects present. Hence, the alternative approach was to develop *statistical models*, based on typical measurements around individual cities, for example. This empirical approach is based on the *log-distance path loss model*, which is an extension of LOS path loss, with the distance exponent of 2 being replaced by a path loss factor n, as given in the following equation:

$$PL(d) = \frac{K}{d^n}, \text{ where } K \text{ is a constant} \tag{7.5}$$

Assuming a reference distance, d_0, which is determined by measurements close to the transmitter in the LOS range, and dividing the two path losses, we obtain the final model as given in the following equation:

$$\frac{PL(d)}{PL(d_0)} = \left(\frac{d_0}{d}\right)^n$$

$$\text{or,} \quad PL(d)_{dB} = PL(d_0)_{dB} + 10n \log\left(\frac{d}{d_0}\right) \tag{7.6}$$

The value of the path loss factor n is usually determined by multiple measurements of path loss within the region of interest, and fitting the measured data to the model in Equation 7.6. Using this approach, typical value of n for *urban cellular radio* falls in the range of 2.7–3.5, while *in-building LOS* has n in the lower range of 1.6–1.8. However, over the years, very specific models have been developed for outdoor propagation (*Hata, Okumura*) and for indoor propagation (*Seidel, Devasirvatham*).

7.4.3 Link Budget and Range Estimation in Wireless Links

For any practical wireless system, it is important to know the maximum reliable data-transmission range. This wireless-system range directly depends on the link budget (L_B) which is defined in the following equation:

$$L_B = P_t + G_t + G_r - R_s \qquad (7.7)$$

where

P_t is the transmitted power in dBm
G_t is the gain of the transmitting antenna in dB
G_r is the gain of the receiving antenna in dB
R_s is the receiver sensitivity in dBm

The *receiver sensitivity*, R_s, is the minimum signal power that the receiver can detect with an acceptable signal-to-noise ratio (SNR), and is analogous to the minimum usable level (MUL) that we defined earlier. If the total path loss between the transmitter and the receiver is greater than the L_B, then communication becomes unreliable. Therefore, it is critical for designers to accurately characterize the path loss and compare it with the L_B to obtain initial estimation of range.

Example

(a) Estimate the transmission range at 900 MHz for an *outdoor cellular channel* ($n = 3$) assuming a system with unity gain transmitting and receiving antennas, a transmitting power of 10 W, and a receiver sensitivity of –100 dBm.
(b) Repeat part (a) for an *indoor LOS channel* ($n = 1.7$), and receiver sensitivity of –20 dBm.

Solution

(a) Using Equation 7.7, with a transmitted power of 10 W or 40 dBm, the system's link budget is calculated as follows:

$$L_B = P_t + G_t + G_r - R_s$$

$$= 40 + 0 + 0 - (-100)$$

$$= 140 \, dBm$$

Now, equating the link budget to the path loss in Equation 7.4 (modified with $n = 3$), we obtain:

$$PL(dB) = L_B = 140$$

$$or, \quad 140 = 30\log(4\pi d) - G_t(dB) - G_r(dB) - 20\log(\lambda)$$

$$= 30\log(4\pi d) - 0 - 0 - 20\log\left(\frac{3 \times 10^8}{900 \times 10^6}\right)$$

Solving, we obtain $d = 1775.73$ m, which is the *cellular* range.

(b) Similarly, for the indoor channel, the link budget is calculated as follows:

$$L_B = P_t + G_t + G_r - R_s$$

$$= 40 + 0 + 0 - (-20)$$

$$= 60 \text{ dBm}$$

and Equation 7.4 is modified for $n = 1.7$, which yields the following final form:

$$PL(\text{dB}) = L_B = 60$$

or, $\quad 60 = 17\log(4\pi d) - G_t(\text{dB}) - G_r(\text{dB}) - 20\log(\lambda)$

$$= 17\log(4\pi d) - 0 - 0 - 20\log\left(\frac{3 \times 10^8}{900 \times 10^6}\right)$$

Solving, we obtain $d = 73.94$ m, which is the *indoor* range.

Problem Solving

P7.1 Give one-word answers to the following specifications of the Bluetooth open standard (refer to Internet sources if required):

(a) Operating frequency band
(b) Type of multiple access
(c) Type of modulation
(d) Type of duplex operation
(e) Channel rate per user
(f) Bit error rate (BER) performance
(g) IEEE specification

P7.2

(a) Give three broad aims of 3G vision.
(b) Give at least one advantage and one disadvantage of each of the following 3G systems: W-CDMA, CDMA2000, and TD-SCDMA.
(c) Select the properties (1G, 2G, 3G, or satellite) that apply to the following systems:

(i) TACS

(ii) Pacific Digital Cellular (PDC)

(iii) cdma2000

(iv) Nordic Mobile Telephony (NMT)

(v) CT2

(vi) Iridium

(vii) Immarsat

P7.3 Find the channel capacity of a W-CDMA system having a center frequency of 1.8 GHz, bandwidth of 100 MHz, and S/N ratio of 20 dB. For a FCC specified UWB system operating from 4 to 9.7 GHz, and assuming the same channel capacity as the W-CDMA system, what is the minimum S/N ratio that the system could handle?

P7.4 Please answer briefly to each of the following questions.

(a) Name three popular 2G systems used in cellular applications.

(b) Name three 2.5 G system technologies that evolved in wireless systems.

(c) For part (b), specify the data rates of the systems.

(d) Name three 3G systems that were approved by the IMT-2000.

(e) For part (d), specify the data rates of the systems.

(f) Name three frequency bands utilized in the WLAN systems.

(g) State the IEEE specifications for WLAN and WPAN systems.

P7.5 Assume an LOS channel with a transmit power of 1 W and 3 dB gain transmitting and receiving antennas.

(a) Compare the path loss at a distance of 2 km, at frequencies of 900 and 2400 MHz. Which frequency would be a better choice for the wireless system?

(b) Estimate the maximum range of the system, assuming a receiver sensitivity of −80 dBm, at both frequencies. Again, which frequency would be a better choice?

Computer Laboratory

C7.1 Develop a Simulink® model to implement the Bluetooth transmitter/ receiver system given in Figure 7.3a and b, using Table 7.1, which

TABLE 7.1

Bluetooth System Components and Equivalent Simulink® Blocks

Bluetooth System Component	Corresponding Simulink Block
1 MHz data ± 1 V	Pulse generator at 1 MHz bit rate
GFSK filter	Analog filter design (low pass) with 500 kHz 3 dB cutoff
16 kHz PN Hop 79 levels (0–78)	Repeating sequence stair
VCO	VCO
Transmitter filter	Analog filter design (band pass)
Channel	Gain (−70 dB)
Receive filter	Analog filter design (band pass)
LO	Sine Wave
B.P. filter at 110.6 MHz	Analog filter design (band pass)
Limiter	Rate Limiter
Quadrature circuit at 110.6 MHz	Analog filter design (band pass)

Note: Use basic circuits to replicate VCO if Simulink blockset is not available; scale down frequencies by a factor of 10 to reduce simulation time.

For VCO, use Figure 3.17 with the signal input replaced by pulse input.

identifies blocks from the Simulink library to represent system components:

(a) Set up the Bluetooth transmitter/receiver simulation design as shown in Figure 7.3.

(b) Run the simulation for the case of a hop between two frequencies, and a propagation path loss (between transmitter output and receiver input) of 70 dB. The two channels suggested are Channel 33 and 51, at frequencies of 2435 and 2453 MHz, respectively. To get the channel frequency in MHz, add 2402 to the channel number.

(c) Plot an overlay of the hop command and the transmitter on/off command in a time range of 0–100 μs.

(d) Plot an overlay of the output of the transmitter's Gaussian filter and the receiver's Gaussian filter in a time range of 0–100 μs.

(e) Plot the two received frequencies after a path loss of 70 dB, in a frequency range of 2400–2480 MHz.

C7.2 Develop a Simulink model to implement the UWB transmitter/receiver system given in Figure 7.4, using a *Pulse Amplitude Modulation (PAM)* system. Table 7.2 shows blocks from the Simulink library to represent UWB system components:

TABLE 7.2

UWB System Components and Equivalent Simulink Blocks

UWB System Component	Corresponding Simulink Block
Data	Uniform random generator
Pulse generator	Pulse generator
Data mixing	Multiplier (for PAM)
Channel	Gain block of −50 dB
LNA	Gain block of 40 dB
Data extract	Gain block of 0 dB (for PAM)
Correlator	Matched filter

Note: Scale down frequencies by a factor of 10 to reduce simulation time.

Appendix A: Synthesized Waveform Generators

A.1 Introduction

The 33600A is Keysight's highest performance function generator, a photograph of which is shown in Figure A.1. With 11 standard waveforms plus pulse and arbitrary waveforms, it is the most frequency stable and lowest distortion function generator in its class. The Keysight 33600A, which is the successor of the HP 33250A, also provides internal AM, FM, and FSK modulation capabilities, sweep and burst operation modes, and a color display. The 33250A provides easy access to standard sine, square, ramp, triangle, and pulse waveforms, plus you can create custom waveforms using the 200 MSa/s, 12 bit, 64k-point arbitrary waveform function.

The variable-edge pulse function gives the user unmatched flexibility for design, verification, and test applications. The 33260A also includes GPIB and RS-232 interfaces standard and IntuiLink software to enable simple generation of custom waveforms. Some of the key features of this equipment are as follows:

- 80 MHz sine and square waveforms
- Ramp, triangle, pulse, noise, and DC waveforms
- 12-bit, 200 MSa/s, 64k-point arbitrary waveforms
- AM, FM, and FSK modulation types
- Linear and logarithmic sweeps and burst operation modes
- Graph mode for visual verification of signal settings
- GPIB and RS-232 interfaces included
- Built-in multiple-unit link for synchronous channels

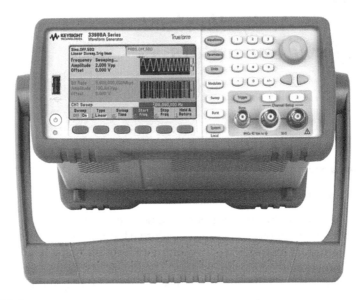

FIGURE A.1
Keysight 33600A waveform generator. (© Keysight Technologies, Inc. 2014. Reproduced with Permission, Courtesy of Keysight Technologies, Santa Rosa, CA.)

A.2 Technical Specifications

A.2.1 Waveforms

- *Standard*

  Sine, square, pulse, ramp, noise, $\sin(x)/x$, exponential rise, exponential fall, cardiac, DC volts

- *Arbitrary*

  Waveform length: 1–64k points

  Amplitude resolution: 12 bits (including sign)

  Repetition rate: 1 µHz to 25 MHz

  Sample rate: 200 MSa/s

  Filter bandwidth: 50 MHz

  Non-vol. memory: Four 64k waveforms

A.2.2 Frequency Characteristics

Sine: 1 µHz to 80 MHz

Square: 1 µHz to 80 MHz

Pulse: 500 µHz to 50 MHz

Arbitrary: 1 μHz to 25 MHz

Ramp: 1 μHz to 1 MHz

White noise: 50 MHz bandwidth

Resolution: 1 μHz; except pulse, 5 digits

Accuracy (1 year): 2 ppm, 18°C–28°C,

3 ppm, 0°C–55°C

A.2.3 Sinewave Spectral Purity

- *Harmonic distortion*

	≤3 Vpp1 (dBc)	≥3 Vpp (dBc)
DC to 1 MHz:	−60	55
1–5 MHz:	−57	−45
5–80 MHz:	−37	−30

- *Total harmonic distortion*
  DC to 20 kHz: < 0.2% + 0.1 mVrms
- *Spurious (nonharmonic)*
  DC to 1 MHz: −60 dBc
  1–20 MHz: −50 dBc
  20–80 MHz: −50 dBc + 6 dBc/octave
- *Phase noise (30 kHz band)*
  10 MHz: <−65 dBc (typical)
  80 MHz: <−47 dBc (typical)

A.2.4 Signal Characteristics

- *Square wave*
  Rise/fall time: <8 ns
  Overshoot: <5%
  Asymmetry: 1% of period + 1 ns

Jitter (rms):	<2 MHz 0.01% + 525 ps	
	≥2 MHz 0.1% + 75 ps	
Duty cycle	≤25 MHz:	20.0%–80.0%
	25–50 MHz:	40.0%–60.0%
	50–80 MHz:	50.0% fixed

- *Pulse*
  Period: 20.00 ns to 2000.0 s
  Pulse width: 8.0 ns to 1999.9 s

Variable edge time: 5.00 ns to 1.00 ms

Overshoot: <5%

Jitter (rms): 100 ppm + 50 ps

- *Ramp*

  Linearity: <0.1% of peak output

  Symmetry: 0.0%–100.0%

- *Arbitrary*

  Minimum edge time: <10 ns

  Linearity: <0.1% of peak output

  Settling time: <50 ns to 0.5% of final value

  Jitter (rms): 30 ppm + 2.5 ns

A.2.5 Output Characteristics

- *Amplitude (into 50 Ω) 10 mVpp to 10 Vpp*

  Accuracy (at 1 kHz, >10 mVpp, Autorange): ±1% of setting ± 1 mVpp

  Flatness (sinewave relative to 1 kHz, Autorange)

    <10 MHz: ±1% (0.1 dB)

    10–50 MHz: ±2% (0.2 dB)

    50–80 MHz: ±5% (0.4 dB)

  Units: Vpp, Vrms, dBm, high and low level

  Resolution: 0.1 mV or 4 digits

- *Offset (into 50 Ω) ± 5 Vpk ac + dc*

  Accuracy: 1% of setting + 2 mV + 0.5% of amplitude

- *Waveform output*

  Impedance: 50 Ω typical (fixed)

             >10 MΩ (output disabled)

  Isolation: 42 Vpk maximum to earth

  Protection: short-circuit protected; overload automatically disables main output

A.2.6 Modulation

- *AM*

  Carrier waveforms: sine, square, ramp, and arbitrary

  Mod. waveforms: sine, square, ramp, noise, and arbitrary

  Mod. frequency: 2 MHz to 20 kHz

Depth: 0.0%–120.0%

Source: internal/external

- *FM*

  Carrier waveforms: sine, square, ramp, and arbitrary

  Mod. Waveforms: sine, square, ramp, noise, and arbitrary

  Mod. Frequency: 2 MHz to 20 kHz

  Deviation range: DC to 80 MHz

  Source: internal/external

- *FSK*

  Carrier waveforms: sine, square, ramp, and arbitrary

  Mod. Waveform: 50% duty cycle square

  Internal rate: 2 mHz to 1 MHz

  Frequency range: 1 μHz to 80 MHz

  Source: internal/external

- *External modulation input*

  Voltage range: ±5 V full scale

  Input impedance: 10 kΩ

  Frequency: DC to 20 kHz

A.2.7 Burst

Waveforms: sine, square, ramp, pulse, arbitrary, and noise

Frequency: 1 μHz to 80 MHz

Burst count: 1–1,000,000 cycles or infinite

Start/stop phase: −360.0° to +360.0°

Internal period: 1 ms to 500 s

Gate source: external trigger

Trigger source: single manual trigger, internal, external trigger

Trigger delay: N-cycle, infinite 0.0 ns to 85.000 s

A.2.8 Sweep

Waveforms: sine, square, ramp, and arbitrary

Type: linear and logarithmic

Direction: up or down

Start F/Stop F: 100 μHz to 80 MHz

Sweep time: 1 ms to 500 s

Trigger: single manual trigger, internal, external trigger

Marker: falling edge of sync signal (programmable)

A.2.9 System Characteristics

- *Configuration times (typical)*
  Function change
  > Standard: 100 ms
  >
  > Pulse: 660 ms
  >
  > Built-in arbitrary: 220 ms

  Frequency change: 20 ms

  Amplitude change: 50 ms

  Offset change: 50 ms

  Select user arbitrary: <900 ms for <16k pts

  Modulation change: <200 ms

- *Arbitrary Download Times GPIB/RS-232 (115 kbps)*

	Arb Length Binary (s)	ASCII Integer (s)	ASCII Real (s)
64k points	48	112	186
16k points	12	28	44
8k points	6	14	22
4k points	3	7	11
2k points	1.5	3.5	5.5

A.2.10 Trigger Characteristics

- *Trigger input*
  Input level: TTL compatible

  Slope: rising or falling, selectable

  Pulse width: >100 ns

  Input impedance: 10 kΩ, DC coupled

  Latency:
  > Burst: <100 ns (typical)
  >
  > Sweep: <10 µs (typical)

  Jitter (rms)
  > Burst: 1 ns; except pulse, 300 ps
  >
  > Sweep: 2.5 µs

- *Trigger output*
  Level: TTL compatible into 50 Ω

Pulse width: >450 ns
Maximum rate: 1 MHz
Fan-out: ≤4 Keysight 33250A's

A.2.11 Clock Reference

- *Phase offset*
  Range: −360° to +360°
  Resolution: 0.001°
- *External reference input*
  Lock range: 10 MHz ± 35 kHz
  Level: 100 mVpp to 5 Vpp
  Impedance: 1 kΩ nominal, ac coupled
  Lock time: <2 s
- *Internal reference output*
  Frequency: 10 MHz
  Level: 632 mVpp (0 dbm), nominal
  Impedance: 50 Ω nominal, ac coupled

A.2.12 Sync Output

Level: TTL compatible into >1 kΩ
Impedance: 50 Ω nominal

A.2.13 General Specifications

Power supply: 100–240 V, 50–60 Hz or 100–127 V, 50–400 Hz
Power consumption: 140 VA
Operating temp: 0°C–55°C
Storage temperature: −30°C to 70°C
Stored states: 4 named user configurations
Power on state: default or last
Interface: IEEE-488 and RS-232 std.
Language: SCPI-1997, IEEE-488.2
Dimensions ($W \times H \times D$)
  Bench top: $254 \times 104 \times 374$ mm
  Rack mount: $213 \times 89 \times 348$ mm
Weight: 4.6 kg

Safety designed to EN61010-1, CSA1010.1, UL-311-1

EMC tested to EN55011, IEC-1326-1

Vibration and shock: MIL-T-28800E, Type III, Class 5

Acoustic noise: 40 dBA

Warm-up time: 1 h

Calibration interval: 1 year

Warranty: 1 year

A.3 Operating Instructions

The front panels of the Keysight 33250A have several operating keys to operate it in either the single frequency or sweep mode operation. The experi-
ments in this book will involve only the single frequency operation, and
hence, only this operation will be discussed here. The following are the
sequence of steps to be performed while operating the Keysight 33250A synthesized function/sweep generators.

- Turn on the instrument by pressing the power switch standby.
  Power is then applied to all of the instrument circuits, and self-tests
  are performed automatically by the instrument.
- Please check that the signal on/off key is *not* lit. It is an important
  precaution to set the specifications of the waveform before changing
  the *signal on/off* key to lit position.
- Press the *function* key to select the waveform that is required. Use
  the arrow keys to move up and down the menu. Examples of waveforms are *sine, square,* and *triangular.* Press *select* after highlighting
  the required waveform.
- Press the *frequency* key to set the frequency of the waveform. For
  example, use the numeric keys to select 1 MHz as the frequency of
  the waveform.
- Press the *amplitude* key to set the peak-to-peak amplitude of the
  waveform. For example, set the amplitude as 1 V.
- Set the *phase* key at 0°, unless other values are specified.
- Set the *dc offset* as 0 V, unless specified otherwise.
- After selecting and setting the values of the waveform, press the *signal on/off* key such that it is in the lit position. The signal can be fed
  out of the BNC connector to the circuit.

Appendix B: RF Spectrum Analyzers

B.1 Introduction

The Keysight CXA signal analyzer (N9000A), the successor to the 8590L series and is shown in Figure B.1, is a versatile, low-cost tool for essential signal characterization. It helps you accelerate product testing and development on multiple levels: cost reduction, throughput, design enhancement, and beyond. It is the successor to the Keysight 8590L, which is also a low-cost, but full-featured, frequency accurate RF spectrum analyzer designed to meet general-purpose measurement needs. The easy-to-use interface provides access to more than 200 built-in functions. Some of the key features of the Keysight 8590L spectrum analyzer are given here:

- Frequency counter: Eliminates the need for a separate frequency counter with the built-in frequency counter, with ±2.1 kHz accuracy at 1 GHz (±7.6 kHz from 0°C to 50°C).

- Multiple resolution bandwidth filters: Optimizes the trade-offs of speed, sensitivity, and the separation of closely spaced signals with the user's choice of 10 resolution bandwidth filters beginning at 1 kHz.

- 145 dB amplitude measurement range: Measures signals directly with −115 to +30 dBm amplitude measurement range.

- One-button measurement routines: Saves time, setup, and training with one button measurement routines such as adjacent channel power, signal bandwidth, and third-order intercept (TOI).

- Phase noise of 105 dBc/Hz at 30 kHz offset: Uncovers small signals close to carriers with an internal phase noise.

- Dual interfaces: Enables user to operate remotely and print directly with the optional dual interfaces that combine either an HP-IB or RS-232 port with a parallel (Centronics) port.

- Built-in tracking generator: Measures the scalar characteristics of your components with the optional built-in tracking generator and the Keysight 85714A scalar measurements personality.

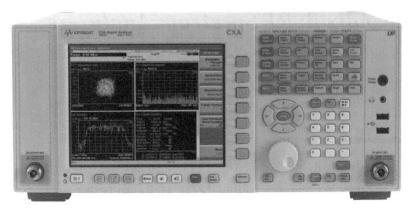

FIGURE B.1
Keysight CXA signal analyzer N9000A. (© Keysight Technologies, Inc. 2014. Reproduced with
Permission, Courtesy of Keysight Technologies, Santa Rosa, CA.)

B.2 Technical Specifications

B.2.1 Frequency Specifications

- *Frequency range*

  50 Ω: 9 kHz to 1.8 GHz

  75 Ω (Opt001): 1 MHz to 1.8 GHz

- *Frequency readout accuracy*

  (Start, stop, center, marker): ±(frequency readout × freq ref
  error + span accuracy + 1% of span + 20% of RBW + 100 Hz)

- *Marker frequency counter accuracy*

  Span 10 MHz: ±(marker freq × freq ref error + counter resolution +
  1 kHz)

- *Counter resolution*

  Span 10 MHz: Selectable from 100 Hz to 100 kHz

- *Frequency span*

  Range: 0 Hz (zero span), 10 kHz to 1.8 GHz

  Resolution: Four digits or 20 Hz, whichever is greater

- Accuracy span 10 MHz: ±3% of span

- *Sweep time*

  Range, Span = 0 Hz or >10 kHz: 20 ms to 100 s

  Accuracy: 20 ms to 100 s: ±3%

  Sweep trigger: Free run, single, line, video, external

B.2.2 Bandwidth Filters

* *Resolution bandwidths*

  1 kHz to 3 MHz (3 dB) in 1, 3, 10 sequence; 9 and 120 kHz (6 dB) EMI bandwidths.

  Accuracy: ±20%

  Selectivity (characteristic) −60 dB/−3 dB:

  > 3–10 kHz: 15:1
  >
  > 100 kHz to 3 MHz: 15:1
  >
  > 1 kHz, 30 kHz: 16:1

* *Video bandwidth range*

  30 Hz to 1 MHz in 1, 3, 10 sequence

* *Stability*

  Noise sidebands (1 kHz RBW, 30 Hz VBW, sample detector)

  >10 kHz offset from CW signal: ≤−90 dBc/Hz

  >20 kHz offset from CW signal: ≤−100 dBc/Hz

  >30 kHz offset from CW signal: ≤−105 dBc/Hz

B.2.3 Amplitude Specifications

* *Measurement range:*

  Displayed average noise level to +30 dBm

  > Opt 001: Displayed average noise level to +75 dBmV

* *Maximum safe input (input attenuator ≥10 dB)*

  Average continuous power: +30 dBm (1 W)

  > Opt 001: +75 dBmV (0.4 W)

  Peak pulse power: +30 dBm (1 W)

  > Opt 001: +75 dBmV (0.4 W)

  DC: 25 Vdc

  > Opt 001: 100 Vdc

* *Gain compression*

  (>10 MHz)

* *Spurious responses*

  Second harmonic distortion:

  5 MHz to 1.8 GHz: <−70 dBc for −45 dBm tone at input mixer

* *Residual responses (input terminated, 0 dB attenuation)*

  150 kHz to 1.8 GHz: <−90 dBm

- *Frequency response (10 dB input attenuation)*

  Absolute (referenced to 300 MHz CAL OUT): ±1.5 dB

  Relative (referenced to midpoint between highest and lowest frequency response deviations): ±1.0 dB

- *Calibrator output*

  Amplitude: −20 dBm ± 0.4 dB

  Opt 001: +28.75 dBmV ± 0.4 dB

- *Resolution bandwidth switching uncertainty (ref to 3 kHz RBW, at ref level)*

  3 kHz to 3 MHz RBW: ±0.4 dB

  1 kHz RBW: ±0.5 dB

- *Log to linear switching*

  0.25 dB at ref level

- *Display scale fidelity*

  Log Incremental accuracy (0 to −60 dB from ref level): ±0.4 dB/4 dB

  Log maximum cumulative (0 to −70 dB from ref level): ±(0.4 + 0.01 × dB from ref level)

  Linear accuracy: ±3% of ref level

B.3 General Specifications

- *Environmental*

  MIL-T-28800: Has been type-tested to the environmental specifications of MIL-T-28800 Class 5

  Temperature

  Operating: 0°C to +55°C

  Storage: −40°C to +75°C

  EMI compatibility

  Conducted and radiated interference

  CISPR Pub.11/1990 Group 1 Class A

  Audible noise: Power requirements On (line 1)

  90–132 Vrms, 47–440 Hz

  195–250 Vrms, 47–66 Hz

- *Power consumption memory*

  User program memory (nominal): 238 kB nonvolatile RAM

- *Data storage (nominal)*
  Internal: 50 traces, 8 states
  External memory cards
  HP 85700A (32 kB), 24 traces or 32 states
  HP 85702A (128 kB), 99 traces or 128 states
- *Video cassette recorder*
  Continuous video recording of display supported through composite video output
- *Size (nominal, without handle, feet, or front cover)*
  325 mm *W* × 163 mm *H* × 427 mm *D*
- *Weight*
  15.2 kg

B.3.1 System Options

- *Option 010 and 011 built-in tracking generators*
  Frequency range
  Opt 010: 100 kHz to 1.8 GHz
  Opt 011: 1 MHz to 1.8 GHz
  Output level range
  Opt 010: −15 to 0 dBm
  Opt 011: +27.8 to +42.8 dBmV
  Resolution: 0.1 dB
  Absolute accuracy at 300 MHz, −10 dBm (+38.8 dBm V, Opt 011): ±1.5 dB
  Vernier range: 15 dB
  Accuracy: ±1 dB
  Output flatness: ±1.75 dB
  Spurious output
  Harmonic spurs: 0 dBm (+ 42.8 dBmV, Opt 011) output
  Dynamic range (characteristic, max output level—TG feed-through)
  Opt 010: 106 dB
  Opt 011: 100 dB
  Power sweep range
  Opt 010: −15 to 0 dBm
  Opt 011: +28.7 to +42.8 dBmV
  Resolution: 0.1 dB

B.3.2 General Options

Opt 003 Memory card reader

Opt 015 Protective soft tan carrying/operating case

Opt 016 Protective soft yellow carrying/operating case

Opt 040 Front panel protective cover with storage and CRT sun shield

Opt 041 HP-IB and parallel printer interfaces

Opt 042 Protective soft carrying case/backpack

Opt 043 RS-232 and parallel printer interfaces

Opt 908 Rack mount kit without handles

Opt 909 Rack mount kit with handles

- *Component test*
  Opt 010 Tracking Generator (100 kHz to 1.8 GHz)
- *Cable TV*
  Opt 001 75 Ω Input
  Opt 011 Tracking generator (75 Ω, 1 MHz to 1.8 GHz)
  Opt 711 50/75 Ω matching pad/100 Vdc block
- *Warranty and support*
  Opt 0Q8 Factory service training
  Opt UK6 Commercial calibration certificate with test data
  Opt AB* Quick reference guide in local languages
  Opt W30 Two additional years return-to-HP service
  Opt W32 Two additional years return-to-HP calibration
  Opt 915 Component level information and service guide
- *Application measurement cards/personalities*
  (Requires Opt 003 memory/measurement card reader)
  HP 85700A Blank 32 kB memory card
  HP 85702A Blank 128 kB memory card
  HP 85714A Scalar measurement personality
  HP 85721A Cable TV measurement personality
  HP 85921A Cable TV PC software for HP 85721A

B.4 Operating Instructions

The Keysight 8590L signal analyzer is a very powerful measuring tool in signal processing and communications. The instrument can very accurately measure the frequency spectrum of input signals in the frequency range of 100 kHz to 1.8 GHz. Basically, the hardware in the instrument calculates the Fourier spectrum of the input signal very rapidly using the fast Fourier transform (FFT).

The front panel of the Keysight 8590L has 3 primary operating keys, which are frequency, span, and amplitude. The following are the sequence of steps to be performed while operating the Keysight 8590L signal analyzer. Please refer also to the Keysight 8590L manual available in the DSP laboratory.

- Turn on the Keysight 8590L by pressing the power switch line. Power is then applied to all of the Keysight 8590L circuits, and self-tests are performed automatically by the instrument.

- Press the *preset* key before starting measurements to automatically calibrate the instrument.

- Press the *frequency* key. Set the desired frequency range in either of the following two ways: Set the *start* and *stop* frequencies, or set the *center* frequency, and enter the *span* of frequency. For example, one could either enter *start* as 0.5 GHz, *stop* as 1.5 GHz, or enter *center* frequency as 1 GHz, with a *span* of 1 GHz.

- The *amplitude* key is used to set the amount of attenuation required on input signal, if it is too strong. *Note*: Signals greater than 30 dBm (Power, dBm = 10 Log (Power, mW)) should not be fed into the instrument; otherwise, it will damage the analyzer.

- If there is an input signal present, then the spectrum of the signal will be displayed on the screen. In order to facilitate measurements, the following controls can be utilized:

  - *Bandwidth (BW)*: Press the *bandwidth* key and lower or increase the resolution bandwidth or the video bandwidth to improve the clarity of the displayed spectrum.

  - *Marker*: Press the *marker* key, and an option menu is displayed. If one presses *peak search*, the instrument will automatically tract the peak value of the spectrum, and position the marker cursor on the peak. One can also press *next peak right* or *next peak left* to move the marker to the adjoining right peak or left peak, respectively.

Appendix C: Dynamic Signal Analyzers

C.1 Introduction

The Keysight 35670A dynamic signal analyzer, shown in Figure C.1, is a versatile dynamic signal analyzer with a built-in source for general spectrum and network analysis and for octave, order, and correlation analyses. Rugged and portable, it is ideal for field work, yet it has the performance and functionality required for demanding R&D applications.

The built-in source, with optional analysis features, optimizes the instrument for analyzing and troubleshooting noise, vibration, and acoustic problems; evaluating and solving rotating machinery problems; and characterizing control systems parameters. Some of the important features of the Keysight 35670A are as follows:

- Frequency range of 102.4 kHz at 1 channel, 51.2 kHz at 2 channel, 25.6 kHz at 4 channel
- 100, 200, 400, 800, and 1600 lines of resolution
- 90 dB dynamic range, 130 dB in swept-sine mode
- Source: Random, burst random, periodic chirp, burst chirp, pink noise, sine, arbitrary waveform
- Measurements: Linear, cross, and power spectrum, power spectral density, frequency response, coherence, THD, harmonic power, time waveform, autocorrelation, crosscorrelation, histogram, PDF, and CDF
- Octave analysis with triggered waterfall display
- Tachometer input and order tracking with orbit diagram
- Built-in 3.5 in. floppy disk

C.2 Technical Specifications

Note: Instrument specifications apply after 15 min warm-up, and within 2 h of the last self-calibration. When the internal cooling fan has been turned

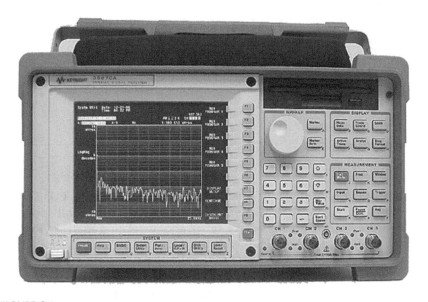

FIGURE C.1
Photograph of the Keysight 35670A dynamic signal analyzer. (© Keysight Technologies, Inc. 2014. Reproduced with Permission, Courtesy of Keysight Technologies, Santa Rosa, CA.)

off, specifications apply within 5 min of the last self-calibration. All specifications are with 400 line frequency resolution and with anti-alias filters enabled unless stated otherwise.

C.2.1 Frequency Specifications

- *Maximum range*

  1 Channel mode: 102.4 kHz, 51.2 kHz (option AY6*)

  2 Channel mode: 51.2 kHz

  4 Channel mode (option AY6 only): 25.6 kHz

- *Spans*

  1 Channel mode: 195.3 MHz to 102.4 kHz

  2 Channel mode: 97.7 MHz to 51.2 kHz

  4 Channel mode (option AY6 only): 97.7 MHz to 25.6 kHz

- *Minimum resolution*

  1 Channel mode: 122 µHz (1600 line display)

  2 Channel mode: 61 µHz (1600 line display)

  4 Channel mode (option AY6 only): 122 µHz (800 line display)

- *Maximum real-time bandwidth*
  (FFT span for continuous data acquisition) (preset, fast averaging)
  1 Channel mode: 25.6 kHz
  2 Channel mode: 12.8 kHz
  4 Channel mode (option AY6 only): 6.4 kHz
- *Measurement rate*
  (Typical) (Preset, fast averaging)
  1 Channel mode: ≥70 averages/s
  2 Channel mode: ≥33 averages/s
  4 Channel mode (option AY6 only): ≥15 averages/s
- *Display update rate*
  Typical (Preset, fast average off): ≥5 updates/s
  Maximum: ≥9 updates/s
  (Preset, fast average off, single channel, single display, undisplayed trace displays set to data registers)
- *Accuracy*
  ±30 ppm (0.003%)

C.2.2 Single-Channel Amplitude

Absolute Amplitude Accuracy (FFT)

(A combination of full scale accuracy, full scale flatness, and amplitude linearity)

±2.92% (0.25 dB) of reading

±0.025% of full scale

- *FFT full scale accuracy at 1 kHz (0 dBfs)*
  ±0.15 dB (1.74%)
- *FFT full scale flatness (0 dBfs)*, Relative to 1 kHz
  ±0.2 dB (2.33%)
- *FFT amplitude linearity at 1 kHz*
  Measured on +27 dBVrms range with time avg
  0 to −80 dBfs ±0.58% (0.05 dB) of reading ±0.025% of full scale
- *Amplitude resolution*
  (16 bits less 2 dB over-range) with averaging 0.0019% of full scale (typical)
- *Residual DC response (FFT mode)*
  Frequency display (excludes A-weight filter)
  <−30 dBfs or 0.5 mVdc

C.2.3 FFT Dynamic Range

- *Spurious free dynamic range*

  90 dB typical (<–80 dBfs)

  (includes spurs, harmonic distortion, intermodulation distortion, alias products)

  Excludes alias responses at extremes of span.

  Source impedance = 50.

  800 line display.

- *Full span FFT noise floor (typical)*

  Flat top window, 64 rms averages, 800 line display.

- *Harmonic distortion*

  <–80 dBfs

  Single tone (in band), ≤0 dBfs

- *Intermodulation distortion*

  <–80 dBfs

  Two tones (in-band), each ≤–6.02 dBfs

- *Spurious and residual responses*

  <–80 dBfs

  Source impedance = 50.

- *Frequency alias responses*

  Single tone (out of displayed range), ≤0 dBfs, ≤1 MHz

  (≤200 kHz with IEPE transducer power supply on)

  2.5%–97.5% of the frequency span: <–80 dBfs

  Lower and upper 2.5% of frequency span: <–65 dBfs

C.2.4 Input Noise

- *Input noise level*

  Flat Top Window, –51 dBVrms range, Source impedance = 50 Ω

  Above 1280 Hz: <–140 dBVrms/$\sqrt[2]{Hz}$

  160 to 1280 Hz: <–130 dBVrms/$\sqrt[2]{Hz}$

  Note: To calculate noise as dB below full scale:

  Noise (dBfs) = Noise (dB/$\sqrt[2]{Hz}$) + 10 log (NBW) – Range (dBVrms); where NBW is the noise equivalent BW of the Window (see the following table).

C.2.5 Window Parameters

Uniform	Hann	Flat	Top
–3 dB bandwidth	0.125% of span	0.185% of span	0.450% of span
Noise equivalent bandwidth	0.125% of span	0.1875% of span	0.4775% of span
Attenuation at ±1/2 Bin	4.0 dB	1.5 dB	0.01 dB
Shape factor (–60 dB BW/–3 dB BW) 800 Hz span	716	9.1	2.6

C.2.6 Single-Channel Phase

- *Phase accuracy relative to external trigger:* ±4.0°

  16 Time Averages Center of Bin, DC coupled

  0 dBfs to –50 dBfs only

  0 Hz < freq 10.24 kHz only

  For Hann and Flat Top windows, phase is relative to a cosine wave at the center of the time record. For the uniform, force, and exponential windows, phase is relative to a cosine wave at the beginning of the time record.

C.2.7 Cross-Channel Amplitude

- *FFT cross-channel gain accuracy:* ±0.04 dB (0.46%)

  Frequency response mode, same amplitude range

  At full scale: Tested with 10 rms Averages on the –11 to +27 dBVrms Ranges, and 100 rms Averages on the –51 dBVrms range

C.2.8 Cross-Channel Phase

- *Cross-channel phase accuracy:* ±0.5°

  (Same conditions as cross-channel amplitude)

C.2.9 Input

- *Input ranges (full scale)*

  (Autorange capability): +27 dBVrms (31.7 Vpk) to – 51 dBVrms (3.99 mVpk) in 2 dB steps
- *Maximum input levels*

  42 Vpk
- *Input impedance*

  1 MΩ ±10%, 90 μF nominal

- *Low side to chassis impedance*
  1 MΩ ± 30% (typical)
  Floating mode: <0.010 μF
  Grounded mode: ≤100 Ω
- *AC coupling roll-off*
  <3 dB roll-off at 1 Hz
  Source impedance = 50 Ω
- *Common mode rejection ratio*
  Single tone at or below 1 kHz
  −51 to −11 dBVrms ranges: >75 dB typical
  − 9 to +9 dBVrms ranges: >60 dB typical
  +11 to +27 dBVrms ranges: >50 dB typical
- *Common mode range (floating mode)*
  ±4 V pk
- *IEPE transducer power supply*
  Current source: 4.25 ± 1.5 mA
  Open circuit voltage: +26 to +32 Vdc
- *A-weight filter, type 0 tolerance*
  Conforms to ANSI Standard S1.4-1983; and to IEC 651-1979;
  10 Hz to 25.6 kHz
- *Crosstalk*
  Between input channels, and source-to-input (receiving channel source impedance = 50 Ω): <−135 dB below signal
  or <−80 dBfs of receiving channel, whichever response is greater in amplitude
- *Time domain*
  Specifications apply in histogram/time mode, and unfiltered time display
- *DC amplitude accuracy*
  ±5.0% fs
- *Rise time of −1 to 0 V test pulse*
  <11.4 μs
- *Settling time of −1 to 0 V test pulse*
  <16 μs to 1%
- *Peak overshoot of −1 to 0 V test pulse*
  <3%

- *Sampling period*
  1 Channel mode: 3.815 µs to 2 s in 2× steps
  2 Channel mode: 7.629 µs to 4 s in 2× steps
  4 Channel mode (Option AY6 Only): 15.26 µs to 8 s in 2× steps

C.2.10 Trigger

- *Trigger modes*
  Internal, source, external (analog setting) GPIB
- *Maximum trigger delay*
  Post trigger: 8191 s
  Pre trigger: 8191 sample periods
  No two channels can be further than ±7168 samples from each other.
- *External trigger max input*
  ±42 Vpk
- *External trigger range*
  Low range: −2 to +2 V
  High range: −10 to +10 V
- *External trigger resolution*
  Low range: 15.7 mV
  High range: 78 mV

C.2.11 Tachometer

- *Pulses per revolution*
  0.5–2048
- *RPM*
  $5 \leq RPM \leq 491,519$
- *RPM accuracy*
  ±100 ppm (0.01%) (typical)
- *Tach level range*
  Low range: −4 to +4 V
  High range: −20 to +20 V
- *Tach level resolution*
  Low range: 39 mV
  High range: 197 mV
- *Maximum tach input level*
  ±42 Vpk

- *Minimum tach pulse width*
  600 ns
- *Maximum tach pulse rate*
  400 kHz (typical)

C.2.12 Source Output

- *Source types*
  Sine, random noise, chrip, pink noise, burst, random, burst chirp
- *Amplitude range*
  AC: ±5 V peak*
  DC: ±10 V*
- *AC amplitude resolution*
  Voltage > 0.2 Vrms: 2.5 mVpeak
  Voltage < 0.2 Vrms: 0.25 mVpeak
- *DC offset accuracy*
  ±15 mV ± 3% of (|DC| + Vacpk) settings
- *Pink noise adder*
  Add 600 mV typical when using pink noise
- *Output impedance*
  <5 Ω
- *Maximum loading*
  Current: ±20 mA peak
  Capacitance: 0.01 µF
- *Sine amplitude accuracy at 1 kHz:* ±4% (0.34 dB) of setting
  Rload > 250 Ω, 0.1–5 Vpk
- *Sine flatness (relative to 1 kHz):* ±1 dB
  0.1–5 V peak
- *Harmonic and subharmonic distortion and spurious signals (in band)*
  0.1–5 Vpk sine wave
  Fundamental < 30 kHz: <–60 dBc
  Fundamental > 30 kHz: <–40 dBc

C.2.13 Digital Interfaces

- *External keyboard*
  Compatible with PC-style 101-key keyboard

---

* Vacpk + |Vdc| ≤10 V.

- *GPIB conforms to the following standards:*
  IEEE 488.1 (SH1, AH1, T6, TE0, L4, LE0, SR1, RL1, PP0, DC1, DT1, C1, C2, C3, C12, E2), IEEE 488.2-1987 Complies with SCPI 1992
- *Data transfer rate*
  <45 ms for a 401 point trace
  (REAL 64 Format)
- *Serial port*
- *Parallel port*
- *External VGA port*

C.2.14 Computed Order Tracking: Option 1D0

Maximum Order × Maximum RPM ≤ 60

- *Online (real time)*
  1 Channel mode: 25,600 Hz
  2 Channel mode: 12,800 Hz
  4 Channel mode: 6,400 Hz
- *Capture playback*
  1 Channel mode: 102,400 Hz
  2 Channel mode: 51,200 Hz
  4 Channel mode: 25,600 Hz
- *Number of orders ≤ 200*
  5 ≤ RPM ≤ 491,519
  (Maximum useable RPM is limited by resolution, tach pulse rate, pulses/revolution, and average mode settings.)
- *Delta order*
  1/128 to 1/1
- *Resolution*
  ≤400
  (Maximum order)/(delta order)
- *Maximum RPM ramp rate:* 1000 RPM/s real-time (typical)
  1,000–10,000 RPM run up
  Maximum order: 10
  Delta order: 0.1
  RPM step: 30 (1 Channel)
  60 (2 Channel)
  120 (4 Channel)

- *Order track amplitude accuracy*
  ±1 dB (typical)

C.2.15 Real-Time Octave Analysis: Option 1D1

- *Standards*

  Conforms to ANSI Standard S1.11-1986, Order 3, Type 1-D, Extended and Optional Frequency Ranges

  Conforms to IEC 651-1979 Type 0 Impulse, and ANSI S1.4

- *Frequency ranges (at centers)*
  Online (Real Time):

	Single Channel (kHz)	2 Channel (kHz)	4 Channel (kHz)
1/1 Octave	0.063–16	0.063–8	0.063–4
1/3 Octave	0.08–40	0.08–20	0.08–10
1/12 Octave	0.0997–12.338	0.0997–6.169	0.0997–3.084
Capture Playback:			
1/1 Octave	0.063–16 kHz	0.063–16 kHz	0.063–16 kHz
1/3 Octave	0.08–31.5 kHz	0.08–31.5 kHz	0.08–31.5 kHz
1/12 Octave	0.0997–49.35 kHz	0.0997–49.35 kHz	0.0997–49.35 kHz

One to 12 octaves can be measured and displayed.

1/1–, 1/3–, and 1/12– octave true center frequencies related by the formula: $f(i + 1)/f(i) = 2^{(1/n)}$; n = 1, 3, or 12; where 1000 Hz is the reference for 1/1, 1/3 Octave, and $1000 \times 2^{(1/24)}$ Hz is the reference for 1/12 octave. The marker returns the ANSI standard preferred frequencies.

- *Accuracy*

  1 se Stable average

  Single tone at band center: ≤±0.20 dB

  Readings are taken from the linear total power spectrum bin. It is derived from sum of each filter.

- *1/3-Octave dynamic range:* >80 dB (typical) per ANSI S1.11-1986

  2 s stable average

  Total power limited by input noise level

C.2.16 Swept-Sine Measurements: Option 1D2

- *Dynamic range:* 130 dB

  Tested with 11 dBVrms

  source level at: 100 ms integration

C.2.17 Arbitrary Waveform Source: Option 1D4

- *Amplitude range*
  AC: ±5 V peak*
  DC: ±10 V*
- *Record length*
  # of Points = 2.56 × Lines of Resolution
  or # of Complex Points = 1.28 × Lines of Resolution
- *DAC resolution*
  0.2828–5 Vpk: 2.5 mV
  0–0.2828 Vpk: 0.25 mV

C.3 General Specifications

- *Safety standards*
  CSA Certified for Electronic Test and Measurement Equipment per CSA C22.2, No. 231
  This product is designed for compliance to: UL1244, Fourth Edition
  IEC 348, 2nd Edition, 1978
- *EMI/RFI standards*
  CISPR 11
- *Acoustic power*
  LpA < 55 dB (cooling fan at high speed setting)
  <45 dB (Auto speed setting at 25°C)
  Fan speed settings of high, automatic, and off are available. The Fan off setting can be enabled for a short period of time, except at higher ambient temperatures where the fan will stay on.
  Environmental operating restrictions

	Operating (Disk in Drive)	Operating (No Disk in Drive)	Storage and Transport
Ambient temperature	4°C–45°C	0°C–55°C	−40°C to 70°C
Relative humidity (noncondensing)			
Minimum	20%	15%	5%
Maximum	80% at 32°C	95% at 40°C	95% at 50°C
			(Continued)

---

* Vacpk + |Vdc| = 10 V.

	Operating (Disk in Drive)	Operating (No Disk in Drive)	Storage and Transport
Vibrations (5–500 Hz)	0.6 Grms	1.5 Grms	3.41 Grms
Shock	5G (10 ms 1/2 sine)		
	5G (10 ms 1/2 sine)		
	40G (3 ms 1/2 sine)		
Max. altitude	4,600 m	4,600 m	4,600 m
	(15,000 ft)	(15,000 ft)	(15,000 ft)

- *AC power*
  90 Vrms–264 Vrms (47–440 Hz)
  350 VA maximum
- *DC power*
  12–28 VDC Nominal
  200 VA maximum
- *DC current at 12 V*
  Standard: <10 A typical
  4 channel: <12 A typical
- *Warm-up time*
  15 min
- *Weight*
  15 kg (33 lb) net
  29 kg (64 lb) shipping
- *Dimensions (excluding bail handle and impact cover)*
  Height: 190 mm (7.5″)
  Width: 340 mm (13.4″)
  Depth: 465 mm (18.3″)

C.4 Operating Instructions

The Keysight 35670A dynamic signal analyzer is a very advanced and versatile signal measuring instrument. The instrument can work in two modes and perform a variety of functions. However, the 35670A can handle baseband signals from 0 to 100 kHz, and has a very high resolution in that range. The two operating modes are single-channel and dual-channel modes. In the single-channel mode, the instrument acts as a time/frequency measuring tool, whereas in the dual-channel mode, the instrument can measure the frequency response of a circuit device, such as a filter.

C.4.1 Single-Channel Mode Operation

- Turn on the 35670A by pressing the power key. Press the *preset* key and press *do preset*, which calibrates the instrument back to its default values.

- Press *inst mode* and press 1 *channel*. Now the instrument is set to perform time/frequency measurements on input signals.

- Connect the input signal to either *channel 1* or *channel 2*.

- Press *meas data*, and select either *time* (Channel 1 or 2), or *spectrum* (*Channel 1 or 2*).

- Press *autoscale*, and the time or frequency graph will be displayed. Use the *marker*, and *marker to peak*, keys to determine the amplitude levels of the signal.

C.4.2 Dual-Channel Mode Operation

- Turn on the 35670A by pressing the power key. Press the *preset* key and press *do preset*, which calibrates the instrument back to its default values.

- Press *inst mode* and press 2 *channel*. Now the instrument is set to perform device frequency response measurements.

- Connect the *source* key of the 35670A to the input port of the device under test (DUT) using BNC cable. Using a BNC Tee, simultaneously connect the *source* key of the 35670A to *channel 1*.

- Connect the output port of the DUT to *channel 2* of the 35670A. Now all connections are complete for making frequency response measurements.

- Press *meas data* and press *frequency response*.

- Press *source* and toggle to *source on/off*. Select any one of the different sources listed, example, and random noise.

- Press *level*, and set the voltage (or power) level of the source.

Press *autoscale*, and the frequency response of the DUT will now be displayed. Use the *marker*, and *marker to peak*, keys to determine the amplitude levels of the signal.

Appendix D: Digital Storage Oscilloscopes

D.1 Introduction

The Keysight DS090254A digital storage oscilloscope, which is the successor to the 54600 series, provides the channel count and measurement power that the user needs, including MegaZoom deep memory, high-definition display, and flexible triggering, especially if designs include heavy analog content. Whether testing is for designs with four inputs, such as antilock brakes, or monitoring multiple outputs of a power supply, the four-channel model helps you get your debug and verification done with ease. Some of the important features of this equipment are as follows:

- Enhanced serial triggering capabilities and integrated five-digit frequency counter measurement
- Lower cost deep memory four-channel scope on the market
- Unique four-channel model
- 350 MHz, 200 MSa/s
- 2 MB of MegaZoom deep memory per channel
- Patented high-definition display
- Flexible triggering including I²C, SPI, CAN, and USB

Figure D.1 shows the photograph of the Keysight 54641D.

D.2 Performance Characteristics of the Keysight 54600 Series Digitizing Oscilloscopes

D.2.1 Acquisition: Analog Channels

- *Max sample rate*
  54621A/D, 54622A/D, 54624A: 200 MSa/s
  54641A/D, 54642A/D: 2 GSa/s interleaved, 1 GSa/s each channel

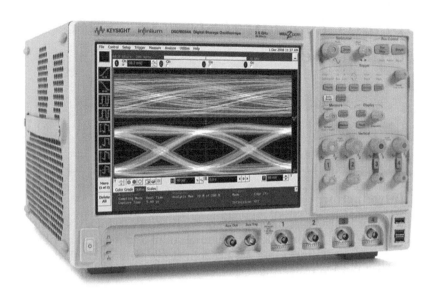

FIGURE D.1
Keysight DS090254A digital storage oscilloscope. (© Keysight Technologies, Inc. 2014. Reproduced with Permission, Courtesy of Keysight Technologies, Santa Rosa, CA.)

- *Max memory depth*
  54621A/D, 54622A/D, 54624A: 4 MB interleaved, 2 MB each channel
  54641A/D, 54642A/D: 8 MB interleaved, 4 MB each channel
- *Vertical resolution*
  8 bits
- *Peak detection*
  54621A/D, 54622A/D, 54624A: 5 ns
  54641A/D, 54642A/D: 1 ns at max sample rate
- *Averaging*
  Selectable from 2, 4, 8, 16, 32, 64, ... to 16,383
- *High-resolution mode*
  54621A/D, 54622A/D, 54624A: 12 bits of resolution when = 500 μs/div, (average mode with ave = 1)
  54641A/D, 54642A/D: 12 bits of resolution when = 100 μs/div, (average mode with ave = 1)
- *Filter*
  Sin x/x interpolation (single shot BW = sample rate/4 or bandwidth of scope, whichever is less) with vectors on

D.2.2 Acquisition: Digital Channels (54621D, 54622D, 54641D, and 54642D Only)

- *Max sample rate*
  54621D, 54622D: 400 MSa/s interleaved, 200 MSa/s each channel
  54641D, 54642D: 1 GSa/s
- *Max memory depth*
  54621D, 54622D: 8 MB interleaved, 4 MB ea. channel
  54641D, 54642D: 4 MB
- *Vertical resolution*
  1 bit
- *Glitch detection (min pulse width)*
  5 ns

D.2.3 Vertical System: Analog Channels

- *Analog channels*
  54621A/D, 54622A/D, 54641A/D, 54642A/D: Ch 1 and 2 simultaneous acquisitions
  54624A: Ch 1, 2, 3, and 4 simultaneous acquisition
- *Bandwidth (−3 dB)**
  54621A/D: dc to 60 MHz
  54622A/D, 54624A: dc to 100 MHz
  54641A/D: dc to 350 MHz
  54642A/D: dc to 500 MHz
- *ac coupled*
  54621A/D: 3.5 Hz to 60 MHz
  54622A/D: 54624A: 3.5 Hz to 100 MHz
  54641A/D: 3.5 Hz to 350 MHz
  54642A/D: 3.5 Hz to 500 MHz
- *Calculated rise time (=0.35/bandwidth)*
  54621A/D: ~5.8 ns
  54622A/D, 54624A: ~3.5 ns
  54641A/D: ~1.0 ns
  54642A/D: ~700 ps

---

* Denotes warranted specifications, all others are typical. Specifications are valid after a 30 min warm-up period and ±10°C from firmware calibration temperature.

- *Single shot bandwidth*

  54621A/D, 54622A/D, 54624A: 50 MHz

  54641A/D: 350 MHz maximum

  54642A/D: 500 MHz maximum

- *Range 1,2*

  54621A/D, 54622A/D, 54624A: 1 mV/div to 5 V/div

  54641A/D, 54642A/D: 2 mV/div to 5 V/div

- *Maximum input*

  CAT I 300 Vrms, 400 Vpk, CAT II 100 Vrms, 400 Vpk

  With 10073C/10074C 10:1 probe: CAT I 500 Vpk, CAT II 400 Vpk

  5 Vrms with 50 Ω input

- *Offset range*

  54621A/D, 54622A/D, 54624A: ±5 V on ranges <10 mV/div; ±25 V on ranges 10 to 199 mV/div; ±100 V on ranges = 200 mV/div

  54641A/D, 54642A/D: ±5 V on ranges <10 mV/div; ±20 V on ranges 10 to 200 mV/div; ±75 V on ranges >200 mV/div

- *Dynamic range*

  Lesser of ±8 div or ±32 V from center screen

- *Input resistance*

  54621A/D, 54622A/D, 54624A: 1 MΩ. ±1%

  54641A/D, 54642A/D: 1 MΩ. ±1%, 50 Ω. selectable

- *Input capacitance*

  54621A/D, 54622A/D, 54624A: ~14 pF

  54641A/D, 54642A/D: ~13 pF

- *Coupling*

  54621A/D, 54622A/D, 54624A: ac, dc, ground

  54641A/D, 54642A/D: ac, dc

- *BW limit*

  54621A/D, 54622A/D, 54624A: ~20 MHz selectable

  54641A/D, 54642A/D: ~25 MHz selectable

- *Channel-to-channel isolation (with channels at same V/div)*

  54621A/D, 54622A/D, 54624A: dc to 20 MHz > 40 dB; 20 MHz to max bandwidth > 30 dB

  54641A/D, 54642A/D: DC to max bandwidth >40 dB

- *Probes*

  54621A/D, 54622A/D, 54624A: 10:1 10074C shipped standard for each analog channel

  54641A/D, 54642A/D: 10:1 10073C shipped standard for each analog channel

- *Probe ID (Keysight/HP and Tek compatible)*

  Auto probe sense

- *ESD tolerance*

  ±2 kV

- *Noise peak-to-peak*

  54621A/D, 54622A/D, 54624A: 2% full scale or 1 mV, whichever is greater

  54641A/D, 54642A/D: 3% full scale or 3 mV, whichever is greater

- *Common mode rejection ratio*

  20 dB at 50 MHz

- *DC vertical gain accuracy*

  ±2.0% full scale

- *DC vertical offset accuracy*

  54621A/D, 54622A/D, 54624A: <200 mV/div: ±0.1 div ±1.0 mV ±0.5% offset value; =200 mV/div: ±0.1 div ±1.0 mV ±1.5% offset value

  54641A/D, 54642A/D: ≤200 mV/div: ±0.1 div ±2.0 mV ±0.5% offset value; >200 mV/div: ±0.1 div ±2.0 mV ±1.5% offset value

- *Single cursor accuracy*

  ±{DC Vertical Gain Accuracy + DC Vertical Offset Accuracy + 0.2% full scale (~1/2 LSB)}

  54621A/D, 54622A/D, 54624A example: for 50 mV signal, scope set to 10 mV/div (80 mV full scale), 5 mV offset, accuracy = ±{2.0% (80 mV) + 0.1 (10 mV) + 1.0 mV + 0.5% (5 mV) + 0.2% (80 mV)} = ±3.78 mV

- *Dual cursor accuracy*

  ±{DC Vertical Gain Accuracy + 0.4% full scale (~1 LSB)}

  Example: for 50 mV signal, scope set to 10 mV/div (80 mV full scale), 5 mV offset, accuracy = ±{2.0% (80 mV) + 0.4% (80 mV)} = ±1.92 mV

D.2.4 Vertical System: Digital Channels (54621D, 54622D, 54641D, and 54642D Only)

- *Number of channels*

  16 Digital—labeled D15–D0

- *Threshold groupings*
  Pod 1: D7–D0
  Pod 2: D15–D8
- *Threshold selections*
  TTL, CMOS, ECL, user definable (selectable by pod)
- *User-defined threshold range*
  ±8.0 V in 10 mV increments
- *Maximum input voltage*
  ±40 V peak CAT I
- *Threshold accuracy**
  ±(100 mV + 3% of threshold setting)
- *Input dynamic range*
  ±10 V about threshold
- *Minimum input voltage swing*
  500 mV peak-to-peak
- *Input capacitance*
  ~8 pF
- *Input resistance*
  100 kΩ. ±2% at probe tip
- *Channel-to-channel skew*
  2 ns typical, 3 ns maximum

D.2.5 Horizontal

- *Range*
  54621A/D, 54622A/D, 54624A: 5 ns/div to 50 s/div
  54641A/D, 54642A/D: 1 ns/div to 50 s/div
- *Resolution*
  54621A/D, 54622A/D, 54624A: 25 ps
  54641A/D, 54642A/D: 2.5 ps
- *Vernier*
  1-2-5 increments when off, ~25 minor increments between major settings when on
- *Reference positions*
  Left, Center, Right

- *Delay range*

  54621A/D, 54622A/D, 54624A: Pretrigger (negative delay): Greater of 1 screen width or 10 ms

    Posttrigger (positive delay): 500 s

  54641A/D, 54642A/D: Pretrigger (negative delay): Greater of 1 screen width or 1 ms

    Posttrigger (positive delay): 500 s

- *Analog delta-t accuracy*

  54621A/D, 54622A/D, 54624A: Same channel*: ±0.01% reading ±0.1% screen width ±40 ps

    Channel-to-channel: ±0.01% reading ±0.1% screen width ±80 ps

  54641A/D, 54642A/D: Same channel*: ±0.005% reading ±0.1% screen width ±20 ps

    Channel-to-channel: ±0.005% reading ±0.1% screen width ±40 ps

  Same Channel Example (54641A/D, 54642A/D): for signal with pulse width of 10 μs, scope set to 5 μs/div (50 μs screen width), delta-t accuracy = ±{0.005% (10 μs) + 0.1% (50 μs) + 20 ps} = 50.52 ns

- *Digital delta-t accuracy*

  54621A/D, 54622A/D, 54624A: (non-Vernier settings)

    Same Channel: ±0.01% reading ±0.1% screen width ±(1 digital sample period, 2.5 or 5 ns based on sample rate of 200/400 MSa/s)

    Channel-to-channel: ±0.01% reading ±0.1% screen width ±(1 digital sample period, 2.5 or 5 ns) ±channel-to-channel skew (2 ns typical, 3 ns maximum)

  54641A/D, 54642A/D:

    Same channel: ±0.005% reading ±0.1% screen width ±(1 digital sample period, 1 ns)

    Channel-to-channel: ±0.005% reading ±0.1% screen width ±(1 digital sample period) ±channel-to-channel skew

  Same channel example (54641A/D, 54642A/D): For signal with pulse width of 10 μs, scope set to 5 μs/div (50 μs screen width), delta-t accuracy = ±{.005%(10 μs) + 0.1% (50 μs) + 1 ns} = 51.5 ns

- *Delay jitter*

  <1 ppm

- *RMS jitter*

  0.025% screen width + 30 ps

- *Modes*

  Main, delayed, roll, XY

- *XY*

  Bandwidth: Maximum bandwidth

  Phase error at 1 MHz: 1.8°

  Z Blanking: 1.4 V blanks trace (use external trigger)—54621A/D, 54622A/D, 54624A only

D.2.6 Trigger System

- *Sources*

  54621A/622A, 54641A/642A: Ch 1, 2, line, ext

  54621D/622D, 54641D/642D: Ch 1, 2, line, ext, D15–D0

  54624A: Ch 1, 2, 3, 4, line, ext

- *Modes*

  Auto, Triggered (normal), Single Auto level (54621A/D, 54622A/D, 54624A only)

- *Hold-off time*

  ~60 ns to 10 s

- *Selections*

  Edge, pulse width, pattern, tv, duration, sequence, CAN, LIN, USB, I2C, SPI

- *Edge*

  Trigger on a rising or falling edge of any source

- *Pattern*

  Trigger on a pattern of high, low, and do not care levels and/or a rising or falling edge established across any of the sources. The analog channel's high or low level is defined by that channel's trigger level.

- *Pulse width*

  Trigger when a positive- or negative-going pulse is less than, greater than, or within a specified range on any of the source channels.

  Minimum pulse width setting: 5 ns (2 ns on 54641A/D, 54642A/D analog channels)

  Maximum pulse width setting: 10 s

- *TV*

  Trigger on any analog channel for NTSC, PAL, PAL-M, or SECAM broadcast standards on either positive or negative composite video signals. Modes supported include Field 1, Field 2, or both, all lines, or any line within a field. Also supports triggering on noninterlaced fields. TV triggers sensitivity: 0.5 division of sync signal.

- *Sequence*

  Arm on event A, trigger on event B, with option to reset on event C or time delay.

- *CAN*

  Trigger on CAN (Controller Area Network) version 2.0A and 2.0B signals. It can trigger on the Start of Frame bit of a data frame, a remote transfer request frame, or an overload frame.

- *LIN*

  Trigger on LIN (Local Interconnect Networking) sync break at beginning of message frame.

- *USB*

  Trigger on USB (Universal Serial Bus) Start of Packet, End of Packet, Reset Complete, Enter Suspend, or Exit Suspend on the differential USB data lines. USB low speed and full speed are supported.

- *I2C*

  Trigger on I2C (Inter-IC bus) serial protocol at a start/stop condition or user-defined frame with address and/or data values. Also trigger on Missing Acknowledge, Restart, EEPROM read, and 10-bit write.

- *SPI*

  Trigger on SPI (Serial Protocol Interface) data pattern during a specific framing period. Support positive and negative chip select framing as well as clock idle framing and user specified number of bits per frame.

- *Duration*

  Trigger on a multichannel pattern whose time duration is less than a value, greater than a value, greater than a time value with a timeout value, or inside or outside of a set of time values.

  Minimum duration setting: 5 ns

  Maximum duration setting: 10 s

- *Autoscale*

  Finds and displays all active analog and digital (for 54621D/54622D/54641D/54642D) channels, sets edge trigger mode on highest numbered channel, sets vertical sensitivity on analog channels and thresholds on digital channels, time base to display ~1.8 periods. Requires minimum voltage >10 mVpp, 0.5% duty cycle, and minimum frequency >50 Hz.

D.2.7 Analog Channel Triggering

- *Range (internal)*
  ±6 div from center screen
- *Sensitivity**
  54621A/D, 54622A/D, 54624A: Greater of 0.35 div or 2.5 mV
  54641A/D, 54642A/D: <10 mV/div: greater of 1 div or 5 mV; =10 mV/div: 0.6 div
- *Coupling*
  AC (~3.5 Hz on 54621A/D, 54622A/D, 54624A. ~10 Hz on 54641A/D, 54642A/D), DC, noise reject, HF reject and LF reject (~50 kHz)

D.2.8 Digital (D15–D0) Channel Triggering (54621D, 54622D, 54641D, and 54642D)

- *Threshold range (used defined)*
  ±8.0 V in 10 mV increments
- *Threshold accuracy**
  ±(100 mV + 3% of threshold setting)
- *Predefined thresholds*
  TTL = 1.4 V, CMOS = 2.5 V, ECL = –1.3 V

D.2.9 External (EXT) Triggering

- *Input resistance*
  54621A/D, 54622A/D, 54624A: 1 M., ±3%
  54641A/D, 54642A/D: 1 M. ±3% or 50.
- *Input capacitance*
- 54621A/D, 54622A/D, 54624A: ~14 pF
- 54641A/D, 54642A/D: ~13pF
- *Maximum input*
  CAT I 300 Vrms, 400 Vpk, CAT II 100 Vrms, 400 Vpk
  With 10073C/10074C 10:1 probe: CAT I 500 Vpk, CAT II 400 Vpk
  5 Vrms with 50 Ω input
- *Range*
  54621A/D, 54622A/D, 54624A: ±10 V
  54641A/D, 54642A/D: DC coupling: trigger level ±8V; AC coup./LFR: AC input minus trig level not to exceed ±8V

- *Sensitivity*

  54621A/D, 54622A/D, 54624A: dc to 25 MHz, <75 mV; 25 MHz to max bandwidth, <150 mV

  54641A/D, 54642A/D: dc to 100 MHz, <100 mV; 100 MHz to max bandwidth, <200 mV

- *Coupling*

  AC (~3.5 Hz), DC, noise reject, HF reject, and LF reject (~50 kHz)

- *Probe ID (Keysight/HP and Tek compatible)*

  Auto probe sense for 54621A/622A/641A/642A

D.2.10 Display System

- *Display*

  7 in. raster monochrome CRT

- *Throughput of analog channels*

  25 million vectors/s per channel with 32 levels of intensity

- *Resolution*

  255 vertical by 1000 horizontal points (waveform area) 32 levels of gray scale

- *Controls*

  Waveform intensity on front panel. Vectors on/off; infinite persistence on/off 8 × 10 grid with continuous intensity control

- *Built-in help system*

  Key-specific help in 11 languages displayed by pressing and holding key or soft key of interest

- *Real time clock*

  Time and date (user settable)

D.2.11 Measurement Features

- *Automatic measurements*

  Measurements are continuously updated. Cursors track current measurement

- *Voltage (analog channels only)*

  Peak-to-Peak, maximum, minimum, average, amplitude, top, base, overshoot, preshoot, RMS (DC)

- *Time*

  Frequency, period, + width, – width, and duty cycle on any channels Rise time, fall time, X at max (time at max volts), X at min (time at min volts), delay, and phase on analog channels only.

- *Counter*

  Built-in 5-digit frequency counter on any channel. Counts up to 125 MHz

  Threshold definition

  Variable by percent and absolute value; 10%, 50%, 90% default for time measurements

- *Cursors*

  Manually or automatically placed readout of horizontal (X, .X, 1/.X) and vertical (Y, .Y)

  Additionally digital or analog channels can be displayed as binary or hex values

- *Waveform math*

  One function of $1 - 2$, 1×2, FFT, differentiate, integrate.

  Source of FFT, differentiate, integrate: analog channels 1 or 2 (or 3 or 4 for 54624A), $1 - 2$, $1 + 2$, 1×2

D.2.12 FFT

- *Points*

  Fixed at 2048 points

- *Source of FFT*

  Analog channels 1 or 2 (or 3 or 4 on 54624A only), $1 + 2$, $1 - 2$, 1×2

- *Window*

  Rectangular, Flattop, Hanning

- *Noise floor*

  -70 to -100 dB depending on averaging

- *Amplitude display*

  In dBV, dBm at 50

- *Frequency resolution*

  0.097656/(time per div.)

- *Maximum frequency*

  102.4/(time per div.)

D.2.13 Storage

- *Save/recall (nonvolatile)*

  54621A/D, 54622A/D, 54624A: 3 setups and traces can be saved and recalled internally

  54641A/D, 54642A/D: 4 setups and traces can be saved and recalled internally

- *Floppy disk*
  3.5″ 1.44 MB double density
  Image formats: TIF, BMP
  Data formats: X and Y (time/voltage) values in CSV format
  Trace/setup formats: Recalled

D.2.14 I/O

- *RS-232 (serial) standard port*
  1 port: XON or DTR; 8 data bits; 1 stop bit; parity = none; 9,600, 19,200, 38,400, 57,600 baud rates (use Keysight 34398A cable)
- *Parallel standard port*
  Printer support
- *Printer compatibility*
  HP DeskJet, LaserJet with HP PCL 3 or greater compatibility
  Black and white at 150×150 dpi; Gray scale at 600×600 dpi
  Epson: black and white at 180×180 dpi
  Seiko thermal DPU-414: black and white
- *Optional GPIB interface module (N2757A)*
  Fully programmable with IEEE488.2 compliance
  Typical GPIB throughput of twenty measurements or twenty 2000-point records per second
- *Optional printer kit*
  The N2727A is a thermal printer kit, including printer power, parallel cable, power cable, and paper.

D.2.15 General Characteristics

- *Physical*
  Size: 32.26 cm wide × 17.27 cm high × 31.75 cm deep (without handle)
  Weight: 6.35 kg (14 lb) on 54621A/D, 54622A/D, 54624A; 6.82 kg (15 lb) on 54641A/D, 54642A/D
- *Probe comp output*
  54621A/D, 54642A/D, 54624A: Frequency ~1.2 kHz; Amplitude ~5 V
  54641A/D, 54642A/D: Frequency ~2 kHz; Amplitude ~5 V
- *Trigger out*
  54621A/D, 54622A/D, 54624A: 0–5 V with 50. Source impedance; delay ~55 ns
  54641A/D, 54642A/D: 0–5 V with 50. Source impedance; delay ~22 ns

- *Printer power*
  7.2–9.2 V, 1 A
- *Kensington lock*
  Connection on rear panel for security

D.2.16 Power Requirements

- *Line voltage range*
  54621A/D, 54622A/D, 54624A: 100–240 VAC ±10%, CAT II, automatic selection
  54641A/D, 54642A/D: 100–240 VAC, 50/60 Hz, CAT II, automatic selection; 100–132 VAC, 440 Hz, CAT II, automatic selection
- *Line frequency*
  54621A/D, 54622A/D, 54624A: 47–440 Hz
  54641A/D, 54642A/D: 50/60 Hz, 100–240 VAC; 440 Hz, 100–132 VAC
- *Power usage*
  54621A/D, 54622A/D, 54624A: 100 W max
  54641A/D, 54642A/D: 110 W max

D.2.17 Environmental Characteristics

- *Ambient temperature*
  Operating −10°C to +55°C; Nonoperating −51°C to +71°C
- *Humidity*
  Operating 95% RH at 40°C for 24 h; Nonoperating 90% RH at 65°C for 24 h
- *Altitude*
  Operating to 4,570 m (15,000 ft); Nonoperating to 15,244 m (50,000 ft)
- *Vibration*
  HP/Keysight class B1 and MIL-PRF-28800F; Class 3 random
- *Shock*
  HP/Keysight class B1 and MIL-PRF-28800F; (operating 30 g, 1/2 sine, 11 ms duration, 3 shocks/axis along major axis. Total of 18 shocks)
- *Pollution degree 2*
  Normally only dry nonconductive pollution occurs. Occasionally a temporary conductivity caused by condensation must be expected.
- *Indoor use only*
  This instrument is rated for indoor use only.

D.2.18 Other information

- *Installation categories*
  CAT I: Mains isolated
  CAT II: Line voltage in appliance and to wall outlet
- *Regulatory information*
  Safety
  IEC 61010-1: 1990+A1: 1992+A2: 1995/EN 61010-1: 1994+A2: 1995
  UL 3111
  CSA-C22.2 No. 1010.1: 1992
- *Supplementary information*
  The product herewith complies with the requirements of the Low Voltage Directive 73/23/EEC and the EMC Directive 89/336/EEC, and carries the CE marking accordingly. The product was tested in a typical configuration with HP/Keysight test systems.

D.3 Operating Instructions

The Keysight 54510A digitizing oscilloscope is a very easy-to-use signal measuring instrument. The instrument can handle input signals from 0 to 100 MHz, and has a very high resolution in that range. There are *four* input channels, which gives it the capability to measure four signals simultaneously. The functional operating steps of the HP 54510A are given as follows:

- Turn on the Keysight 54510 by pressing the power key at the *back* of the instrument.
- Connect the input signal to any of the four available channels: *Channel 1, Channel 2, Channel 3, or Channel 4.*
- Press *Autoscale*, and observe the waveform on the display screen.
- To display the *peak-to-peak amplitude* of the signal, press *function key (blue key)*, and press *Vp-p*.
- To display the *maximum amplitude* of the signal, press *function key (blue key)*, and press *Vmax*.
- To display the *minimum amplitude* of the signal, press *function key (blue key)*, and press *Vmin*.
- To display the *frequency* of the signal (if the signal is periodic), press *function key (blue key)* and press *Freq*.

- To display the *time period* of the signal (if the signal is periodic), press *function key (blue key)* and press *Period*.
- Press *Save* to save the currently displayed waveform *to* memory, with a file name.
- Press *Recall* to display any saved waveform *from* memory, by specifying the appropriate file name.
- Press *Display* to adjust the format of the display, such as *gridlines* (*on* or *off*) and *dotted line*, or *full line* display.

Appendix E: Integrated Circuits for Communication Systems

E.1 Introduction to Radio-Frequency Integrated Circuits

In the electronics industry, radio frequency integrated circuits (RFIC) is a generic term used for integrated circuits designed for wireless communication. The frequency range of radiofrequency (RF) normally falls between 300 MHz and 30 GHz. RFIC applications include mobile phone, WLAN, UWB, GPS, and Bluetooth. RF building blocks that are in high demand for integrated transceiver design are as follows:

- Low-noise amplifiers
- Mixers
- Clock generators
- Power amplifiers
- Antenna
- Duplexer
- Passive components (inductor, capacitor, and transformer)

The next sections will give typical examples of modern RFICs that are used currently in the communications industry. Three important categories of devices have been illustrated:

- RF amplifiers
- RF mixers
- RF transceivers

E.2 RFIC Amplifier: TRF37B73 1–6000 MHz RF Gain Block

The TRF37B73, as shown in Figure E.1, is packaged in a 2.00 mm × 2.00 mm WSON with a power down pin feature, making it ideal for applications where space and low-power modes.

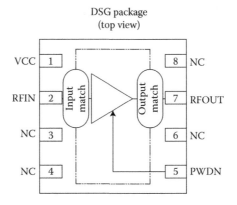

FIGURE E.1
Chip layout of TRF37B73 RF gain block. (Reproduced with permission from ©Texas Instruments.)

Features

- 1–6000 MHz
- Gain: 14.5 dB
- Noise figure: 4 dB
- Output P1dB: 15 dBm at 2000 MHz
- Output IP3: 27 dBm at 2000 MHz
- Power-down mode
- Single supply: 3.3 V
- Stabilized performance over temperature
- Unconditionally stable
- Robust ESD: >1 kV HBM; >1 kV CDM

Applications

- General-purpose RF gain block minimizes needed PCB area.
- Consumer
- Industrial
- Utility meters
- Low-cost radios
- Cellular base station
- Wireless infrastructure
- RF backhaul
- Radar
- Electronic warfare

- Software-defined radio
- Test and measurement
- Point-to-point/multipoint microwave
- Software-defined radios
- RF repeaters
- Distributed antenna systems
- LO and PA driver amplifier
- Wireless Data, satellite, DBS, CATV
- IF amplifier

E.3 RFIC Mixer: TRF37B32 700–2700 MHz Dual Downconverter

The TRF37B32, as shown in Figure E.2, is a wideband dual downconverter mixer with integrated IF amplifier.

Features

- Device family supports wide RF input range
  - TRF37A32: 400–1700 MHz
  - TRF37B32: 700–2700 MHz
  - TRF37C32: 1700–3800 MHz
- Gain: 10 dB
- Noise figure: 9.5 dB

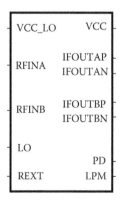

FIGURE E.2
Chip layout of TRF37B32 dual downconverter. (Reproduced with permission from ©Texas Instruments.)

- Input IP3: 30 dBm
- 500 mW per channel power dissipation
- Single-ended RF input
- IF frequency range from 30 to 600 MHz
- 45 dB isolation between channels
- Low-power mode option
- Independent power-down control
- Single 3.3 V supply
- No external matching required

Applications
- Wireless infrastructure
  - WCDMA, TD-SCDMA
  - LTE, TD-LTE
  - Multicarrier GSM (MC-GSM)
- Point-to-point microwave
- Software-defined radios (SDR)
- Radar receiver
- Satellite communications

E.4 RFICs for Transceiver Applications

E.4.1 Texas Instruments CC2540 2.4 GHz Bluetooth® Transceiver

The CC2540, shown in Figure E.3, is a cost-effective, low-power, true system-on-chip (SoC) for *Bluetooth* low-energy applications.

Features	Applications
• True single-chip BLE solution: CC2540 can run both application and BLE protocol stack includes peripherals to interface with wide range of sensors, etc.	• 2.4 GHz Bluetooth low-energy systems • Mobile phone accessories
• 6 mm × 6 mm Package QFN40 Package	• Sports and leisure equipment
• RF: Bluetooth low-energy technology compatible, excellent link budget (up to 97 dB) enabling long-range applications without external front end	• Consumer electronics • Human interface devices (keyboard, mouse, remote control)

(Continued)

Features	Applications
• Low power: Active mode RX down to 19.6 mA, active mode TX (−6 dBm): 24 mA, wide supply voltage range (2–3.6 V), Full RAM and register retention in all power modes	• USB dongles • Health care and medical
• Microcontroller: High-performance and low-power 8051 MCU Core, 8 kB SRAM	
• Peripherals: 12-Bit ADC with eight channels and configurable resolution, integrated high-performance Op-Amp and ultralow-power comparator, general-purpose timers (one 16-Bit, two 8-Bit), 21 general-purpose I/O pins (19 × 4 mA, 2 × 20 mA)	

E.4.2 Texas Instruments TRF2443 Integrated IF Transceiver

The TRF2443, as shown in Figure E.4, is a highly integrated full-duplex intermediate frequency (IF) transceiver designed for broadband point-to-point wireless communications applications.

Features

- Integrated TX chain (165–175 MHz/330–350 MHz)
  - Baseband amplifiers
  - Quadrature modulator
  - Digitally controlled VGA
  - TX output IP3: 29.5 dBm
  - TX output noise: −166 dBc/Hz
- Integrated RX chain (140–165 MHz/280–330 MHz)
  - IF amplifiers
  - Analog and digital VGA
  - Quadrature demodulator
  - Baseband filters
  - ADC buffers
  - IF SAW filter bypass
  - RX noise figure: 4.3 dB
  - RX input IP3: 9.5 dBm
- Integrated TX and RX synthesizers
- Integrated cross-polarization interference Cancellation (XPIC) support
- Auxiliary RX chain

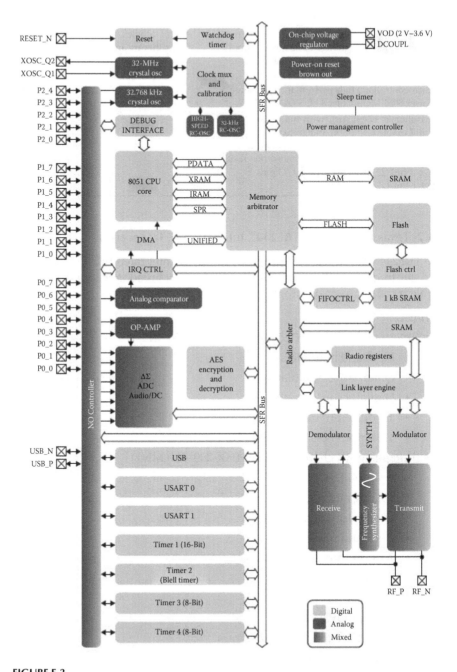

FIGURE E.3
Chip layout of TI CC2540 Bluetooth transceiver. (Reproduced with permission from ©Texas Instruments.)

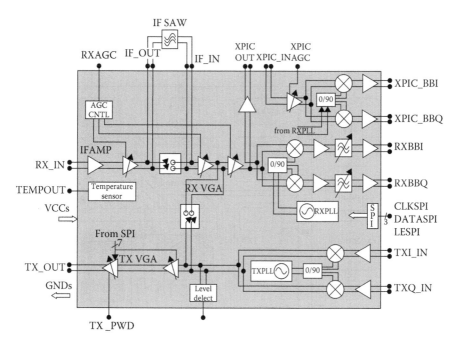

FIGURE E.4
Chip layout of TI TRF2442 IF transceiver. (Reproduced with permission from ©Texas Instruments.)

Applications

- Wireless microwave backhaul
- Point-to-point microwave
- Broadband wireless applications
- WiMAX IF transceiver

E.4.3 Texas Instruments CC2520 ZigBee® RF Transceiver (Figure E.5)

Applications

- IEEE 802.15.4 systems
- ZigBee systems
- Industrial monitoring and control
- Home and building automation
- Automatic meter reading
- Low-power wireless sensor networks
- Set-top boxes and remote controls
- Consumer electronics

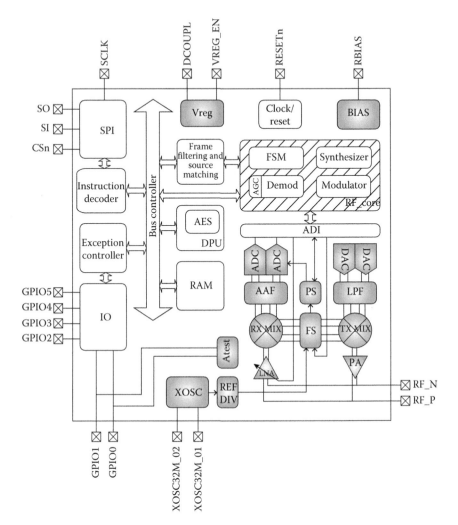

FIGURE E.5
Texas Instruments CC2520 Zigbee RF transceiver. (Reproduced with permission from ©Texas Instruments.)

Features

- State-of-the-art selectivity/co-existence
  Adjacent channel rejection: 49 dB
  Alternate channel rejection: 54 dB
- Excellent link budget (103 dB)
  400 m line-of-sight range
- Extended temperature range (−40°C to +125°C)

- Wide supply range: 1.8–3.8 V
- Extensive IEEE 802.15.4 MAC hardware support to offload the microcontroller
- AES-128 security module
- CC2420 interface compatibility mode

Low power
- RX (receiving frame, −50 dBm) 18.5 mA
- TX 33.6 mA at +5 dBm
- TX 25.8 mA at 0 dBm
- <1 μA in power down

General
- Clock output for single-crystal systems
- RoHS compliant 5 × 5 mm QFN28 (RHD) package

Radio
- IEEE 802.15.4 compliant DSSS baseband modem with 250 kbps data rate
- Excellent receiver sensitivity (−98 dBm)
- Programmable output power up to +5 dBm
- RF frequency range 2394–2507 MHz
- Suitable for systems targeting compliance with worldwide radio frequency regulations: ETSI EN 300 328 and EN 300 440 class 2 (Europe), FCC CFR47 Part 15 (United States) and ARIB STD-T66 (Japan)

Microcontroller support
- Digital RSSI/LQI support
- Automatic clear channel assessment for CSMA/CA
- Automatic CRC
- 768 bytes RAM for flexible buffering and security processing
- Fully supported MAC security
- Four wire SPI
- Six configurable IO pins
- Interrupt generator
- Frame filtering and processing engine
- Random number generator

Development tools

- Reference design
- IEEE 802.15.4 MAC software
- ZigBee stack software
- Fully equipped development kit
- Packet sniffer support in hardware

Appendix F: Worldwide Frequency Bands and Terminology

The International Telecommunication Union (ITU) divides the world into three radio regulatory regions for the purpose of managing and harmonizing global radio spectrum allocation and use. Each region has its own set of frequency allocations, which are generally very similar. The three ITU regions are shown in Figure F.1, and the geographical areas covered in the three regions are as follows:

Region 1
Europe, Middle East, Africa, the former Soviet Union, including Siberia, and Mongolia

Region 2
North and South America and Pacific (East of the International Date Line)

Region 3
Asia, Australia, and the Pacific Rim (West of the International Date Line).

The complete spectrum allocations are given in Table F.1, starting on the next page.

FIGURE F.1
ITU radio spectrum regions. (Reproduced from International Telecommunication Union (ITU) sources.)

TABLE F.1

Worldwide Frequency Allocations

Table of Frequency Allocations — 0–160 kHz (VLF/LF)

International Table			United States Table		FCC Rule Part(s)
Region 1 Table	Region 2 Table	Region 3 Table	Federal Table	Non-Federal Table	
Below 9 (Not Allocated) 5.53 5.54			Below 9 (Not Allocated) 5.53 5.54		
9–14 RADIONAVIGATION			9–14 RADIONAVIGATION US18 US2		
14–19.95 FIXED MARITIME MOBILE 5.57			14–19.95 FIXED MARITIME MOBILE 5.57	14–19.95 Fixed	
5.55 5.56			US2	US2	
19.95–20.05 STANDARD FREQUENCY AND TIME SIGNAL (20 kHz)			19.95–20.05 STANDARD FREQUENCY AND TIME SIGNAL (20 kHz) US2		
20.05–70 FIXED MARITIME MOBILE 5.57			20.05–59 FIXED MARITIME MOBILE 5.57 US2	20.05–59 FIXED US2	
			59–61 STANDARD FREQUENCY AND TIME SIGNAL (60 kHz) US2		

(Continued)

TABLE F.1 (*Continued*)

Worldwide Frequency Allocations

Table of Frequency Allocations   0–160 kHz (VLF/LF)

International Table			United States Table		FCC Rule Part(s)
Region 1 Table	Region 2 Table	Region 3 Table	Federal Table	Non-Federal Table	
5.56 5.58			61–70 FIXED MARITIME MOBILE 5.57 US2	61–70 FIXED US2	
70–72 RADIO- NAVIGATION 5.60	70–90 FIXED MARITIME MOBILE 5.57 MARITIME RADIO- NAVIGATION 5.60 Radiolocation	70–72 RADIO- NAVIGATION 5.60 Fixed Maritime mobile 5.57 5.59	70–90 FIXED MARITIME MOBILE 5.57 Radiolocation	70–90 FIXED Radiolocation	Private Land Mobile (90)
72–84 FIXED MARITIME MOBILE 5.57 RADION- AVIGATION 5.60 5.56		72–84 FIXED MARITIME MOBILE 5.57 RADION AVIGATION 5.60			

(Continued)

TABLE F.1 (*Continued*)

Worldwide Frequency Allocations

Table of Frequency Allocations

0–160 kHz (VLF/LF)

International Table			United States Table		FCC Rule Part(s)
Region 1 Table	Region 2 Table	Region 3 Table	Federal Table	Non-Federal Table	
84–86 RADIONAVIGATION 5.60		84–86 RADIO-NAVIGATION 5.60 Fixed Maritime mobile 5.57 5.59			
86–90 FIXED MARITIME MOBILE 5.57 RADIONAVIGATION 5.56	5.61	86–90 FIXED MARITIME MOBILE 5.57 RADIO-NAVIGATION 5.60	US2	US2	
90–110 RADIONAVIGATION 5.62 Fixed 5.64			90–110 US2 US104	RADIONAVIGATION 5.62 US18	Aviation (87) Private Land Mobile (90)
110–112 FIXED MARITIME MOBILE RADIONAVIGATION 5.64	110–130 FIXED MARITIME MOBILE MARITIME RADIO-NAVIGATION 5.60 Radiolocation	110–112 FIXED MARITIME MOBILE RADIO-NAVIGATION 5.60 5.64	110–130 FIXED MARITIME MOBILE Radiolocation		Private Land Mobile (90)

(Continued)

TABLE F.1 (*Continued*)

Worldwide Frequency Allocations

Table of Frequency Allocations

0–160 kHz (VLF/LF)

International Table			United States Table		FCC Rule Part(s)
Region 1 Table	Region 2 Table	Region 3 Table	Federal Table	Non-Federal Table	
112–115 RADIO- NAVIGATION 5.60		112–117.6 RADIO- NAVIGATION 5.60 Fixed Maritime mobile			
115–117.6 RADIO- NAVIGATION 5.60 Fixed Maritime mobile 5.64 5.66		5.64 5.65			
117.6–126 FIXED MARITIME MOBILE RADIONAVIGATION 5.60 5.64		117.6–126 FIXED MARITIME MOBILE RADIO- NAVIGATION 5.60 5.64			
126–129 RADIONAVIGATION 5.60		126–129 RADIO- NAVIGATION 5.60 Fixed Maritime mobile 5.64 5.65			

(*Continued*)

TABLE F.1 (*Continued*)

Worldwide Frequency Allocations

Table of Frequency Allocations			0–160 kHz (VLF/LF)			
International Table			United States Table			
Region 1 Table	Region 2 Table	Region 3 Table	Federal Table	Non-Federal Table	FCC Rule Part(s)	
129–130 FIXED MARITIME MOBILE RADIO- NAVIGATION 5.60 5.64	5.61 5.64	129–130 FIXED MARITIME MOBILE RADIO- NAVIGATION 5.60 5.64	5.64 US2		Maritime (80)	
130–135.7 FIXED MARITIME MOBILE 5.64 5.67	130–135.7 FIXED MARITIME MOBILE 5.64	130–135.7 FIXED MARITIME MOBILE RADIO- NAVIGATION 5.64	130–160 FIXED MARITIME MOBILE			
135.7–137.8 FIXED MARITIME MOBILE Amateur 5.67A 5.64 5.67 5.67B	135.7–137.8 FIXED MARITIME MOBILE Amateur 5.67A 5.64	135.7–137.8 FIXED MARITIME MOBILE RADIO- NAVIGATION 5.67A Amateur 5.67A 5.64 5.67B				

(Continued)

TABLE F.1 (Continued)

Worldwide Frequency Allocations

Table of Frequency Allocations

0–160 kHz (VLF/LF)

International Table			United States Table		FCC Rule Part(s)
Region 1 Table	**Region 2 Table**	**Region 3 Table**	**Federal Table**	**Non-Federal Table**	
137.8–148.5 FIXED MARITIME MOBILE	137.8–160 FIXED MARITIME MOBILE	137.8–160 FIXED MARITIME MOBILE RADIO-NAVIGATION			
5.64 5.67	5.64	5.64	5.64 US2		

Table of Frequency Allocations

160–1,800 kHz (LF/MF)

International Table			United States Table		FCC Rule Part(s)
Region 1 Table	**Region 2 Table**	**Region 3 Table**	**Federal Table**	**Non-Federal Table**	
148.5–255 BROADCASTING 5.68 5.69 5.70	160–190 FIXED	160–190 FIXED Aeronautical radionavigation	160–190 FIXED MARITIME MOBILE US2	160–190 FIXED US2	
	190–200 AERONAUTICAL RADIONAVIGATION		190–200 AERONAUTICAL RADIO-NAVIGATION US18 US2		Aviation (87)
255–283.5 BROADCASTING AERONAUTICAL RADION-AVIGATION 5.70 5.71	200–275 AERONAUTICAL RADIO-NAVIGATION Aeronautical mobile	200–285 AERONAUTICAL RADIO-NAVIGATION Aeronautical mobile	200–275 AERONAUTICAL RADIONAVIGATION US18 Aeronautical mobile US2	AERONAUTICAL RADIONAVIGATION	

(Continued)

TABLE F.1 (*Continued*)

Worldwide Frequency Allocations

Table of Frequency Allocations — 160–1,800 kHz (LF/MF)

International Table			United States Table		FCC Rule Part(s)
Region 1 Table	Region 2 Table	Region 3 Table	Federal Table	Non-Federal Table	
283.5–315 AERONAUTICAL RADIONAVIGATION MARITIME RADIONAVIGATION (radiobeacons) 5.73	275–285 AERONAUTICAL RADIO-NAVIGATION Aeronautical mobile Maritime radio-navigation (radiobeacons)		275–285 AERONAUTICAL RADIO-NAVIGATION Aeronautical mobile Maritime radionavigation (radiobeacons) US2 US18		
	285–315 AERONAUTICAL RADIONAVIGATION MARITIME RADIONAVIGATION (radiobeacons) 5.73		285–325 MARITIME RADIONAVIGATION (radiobeacons) 5.73 Aeronautical radionavigation (radiobeacons)	MARITIME RADIONAVIGATION Aeronautical radionavigation (radiobeacons)	
5.72 5.74					
315–325 AERONAUTICAL RADIONAVIGATION Maritime radionavigation (radiobeacons) 5.73 5.72 5.75	315–325 MARITIME RADIO-NAVIGATION (radiobeacons) 5.73 Aeronautical radionavigation	315–325 AERONAUTICAL RADIO-NAVIGATION MARITIME RADIO-NAVIGATION (radiobeacons) 5.73	US2 US18 US364		

(*Continued*)

TABLE F.1 (*Continued*)

Worldwide Frequency Allocations

Table of Frequency Allocations — 160–1,800 kHz (LF/MF)

International Table			United States Table		FCC Rule Part(s)
Region 1 Table	Region 2 Table	Region 3 Table	Federal Table	Non-Federal Table	
325–405 AERONAUTICAL RADIONAVIGATION	325–335 AERONAUTICAL RADIO-NAVIGATION Aeronautical mobile Maritime radionavigation (radiobeacons)	325–405 AERONAUTICAL RADIO-NAVIGATION Aeronautical mobile	325–335 AERONAUTICAL RADIONAVIGATION (radiobeacons) Aeronautical mobile Maritime radionavigation (radiobeacons) US2 US18		Aviation (87)
	335–405 AERONAUTICAL RADIO-NAVIGATION Aeronautical mobile		335–405 AERONAUTICAL RADIONAVIGATION (radiobeacons) US18 Aeronautical mobile US2		
5.72					
405–415 RADIONAVIGATION 5.76 5.72	405–415 RADIONAVIGATION Aeronautical mobile		405–415 RADIONAVIGATION 5.76 US18 Aeronautical mobile US2		Maritime (80) Aviation (87)

(Continued)

TABLE F.1 (*Continued*)

Worldwide Frequency Allocations

Table of Frequency Allocations

0–160 kHz (VLF/LF)

International Table			United States Table		FCC Rule Part(s)
Region 1 Table	**Region 2 Table**	**Region 3 Table**	**Federal Table**	**Non-Federal Table**	
415–435 MARITIME MOBILE 5.79 AERONAUTICAL RADIONAVIGATION 5.72	415–495 MARITIME MOBILE 5.79 5.79A Aeronautical radionavigation 5.80		415–435 MARITIME MOBILE 5.79 AERONAUTICAL RADIONAVIGATION US2	415–495 MARITIME MOBILE 5.79 AERONAUTICAL RADIONAVIGATION	
435–495 MARITIME MOBILE 5.79 5.79A Aeronautical radionavigation			435–495 MARITIME MOBILE 5.79 5.79A Aeronautical radionavigation 5.82 US2 US231	435–495 MARITIME MOBILE 5.79 5.79A 5.82 US2 US231	
5.72 5.82	5.77 5.78 5.82				
495–505 MOBILE 5.82A 5.82B			495–505 MOBILE (distress and calling)		

(Continued)

TABLE F.1 (*Continued*)

Worldwide Frequency Allocations

Table of Frequency Allocations

0–160 kHz (VLF/LF)

International Table			United States Table		FCC Rule Part(s)
Region 1 Table	Region 2 Table	Region 3 Table	Federal Table	Non-Federal Table	
505–526.5 MARITIME MOBILE 5.79 5.79A 5.84 AERONAUTICAL RADIONAVIGATION 5.72	505–510 MARITIME MOBILE 5.79	505–526.5 MARITIME MOBILE 5.79 5.79A 5.84 AERONAUTICAL RADIO-NAVIGATION Aeronautical mobile Land mobile	505–510 MARITIME MOBILE 5.79		Maritime (80)
	510–525 MOBILE 5.79A 5.84 AERONAUTICAL RADIO-NAVIGATION		510–525 MARITIME MOBILE (ships only) 5.79A 5.84 AERONAUTICAL RADIONAVIGATION (radiobeacons) US18 US14 US225	MARITIME MOBILE (ships only) 5.79A 5.84 AERONAUTICAL RADIONAVIGATION (radiobeacons) US18	Maritime (80) Aviation (87)
526.5–1606.5 BROADCASTING	525–535 BROADCASTING 5.86 AERONAUTICAL RADIO-NAVIGATION	526.5–535 BROADCASTING Mobile 5.88	525–535 MOBILE US221 AERONAUTICAL RADIONAVIGATION (radiobeacons) US18 US239	526.5–535 BROADCASTING	Aviation (87) Private Land Mobile (90)
	535–1605 BROADCASTING	535–1606.5 BROADCASTING	535–1605	535–1605 BROADCASTING NG1 NG5	Radio Broadcast (AM) (73) Private Land Mobile (90)

(*Continued*)

TABLE F.1 (Continued)

Worldwide Frequency Allocations

Table of Frequency Allocations — 0–160 kHz (VLF/LF)

	International Table			United States Table		FCC Rule Part(s)
	Region 1 Table	Region 2 Table	Region 3 Table	Federal Table	Non-Federal Table	
5.87 5.87A						
1606.5–1625 FIXED MARITIME MOBILE 5.90 LAND MOBILE	1605–1625 BROADCASTING 5.89	1606.5–1800 FIXED MOBILE RADIOLOCATION RADIO-NAVIGATION	1605–1615 MOBILE US221 G127 / 1615–1705	1605–1705 BROADCASTING 5.89	Radio Broadcast (AM) (73) Alaska Fixed (80) Private Land Mobile (90)	
5.92	5.90					
1625–1635 RADIOLOCATION	1625–1705 FIXED MOBILE BROADCASTING 5.89					
5.93	Radiolocation					
1635–1800 FIXED MARITIME MOBILE 5.90 LAND MOBILE	5.90		US299	US299 NG1 NG5		
5.92 5.96	1705–1800 FIXED MOBILE RADIOLOCATION AERONAUTICAL RADIO-NAVIGATION	5.91	1705–1800 FIXED MOBILE RADIOLOCATION US240			Alaska Fixed (80) Private Land Mobile (90)

(Continued)

TABLE F.1 (Continued)

Worldwide Frequency Allocations

Table of Frequency Allocations · 1,800–3,025 kHz (MF/HF)

International Table			United States Table		FCC Rule Part(s)
Region 1 Table	Region 2 Table	Region 3 Table	Federal Table	Non-Federal Table	
1800–1810 RADIOLOCATION 5.93	1800–1850 AMATEUR	1800–2000 AMATEUR FIXED MOBILE except aeronautical mobile RADIO-NAVIGATION Radiolocation	1800–1900	1800–1900 AMATEUR	Amateur Radio (97)
1810–1850 AMATEUR 5.98 5.99 5.100 5.101					
1850–2000 FIXED MOBILE except aeronautical mobile 5.92 5.96 5.103	1850–2000 AMATEUR FIXED MOBILE except aeronautical mobile RADIOLOCATION RADIO-NAVIGATION 5.102	5.97	1900–2000 RADIOLOCATION US290		Private Land Mobile (90) Amateur Radio (97)
2000–2025 FIXED MOBILE except aeronautical mobile (R) 5.92 5.103	2000–2065 FIXED MOBILE		2000–2065 FIXED MOBILE	2000–2065 MARITIME MOBILE	Maritime (80) Private Land Mobile (90)

(Continued)

TABLE F.1 (*Continued*)

Worldwide Frequency Allocations

Table of Frequency Allocations			1,800–3,025 kHz (MF/HF)		
International Table			United States Table		
Region 1 Table	Region 2 Table	Region 3 Table	Federal Table	Non-Federal Table	FCC Rule Part(s)
2025–2045 FIXED MOBILE except aeronautical mobile (R) Meteorological aids 5.104 5.92 5.103					
2045–2160 FIXED MARITIME MOBILE LAND MOBILE 5.92	2065–2107 MARITIME MOBILE 5.105 5.106		US340 2065–2107 MARITIME MOBILE 5.105 US296 US340	US340 NG7	Maritime (80)
2160–2170 RADIOLOCATION	2107–2170 FIXED MOBILE		2107–2170 FIXED MOBILE	2107–2170 FIXED MOBILE except aeronautical mobile	Maritime (80) Private Land Mobile (90)
5.93 5.107			US340	US340 NG7	
2170–2173.5 MARITIME MOBILE			2170–2173.5 MARITIME MOBILE (telephony) US340	2170–2173.5 MARITIME MOBILE US340	Maritime (80)

(*Continued*)

TABLE F.1 (Continued)

Worldwide Frequency Allocations

Table of Frequency Allocations   1,800–3,025 kHz (MF/HF)

International Table			United States Table		FCC Rule Part(s)
Region 1 Table	Region 2 Table	Region 3 Table	Federal Table	Non-Federal Table	
2173.5–2190.5 MOBILE (distress and calling)			2173.5–2190.5 MOBILE (distress and calling)	(distress and calling)	Maritime (80) Aviation (87)
5.108 5.109 5.110 5.111			5.108 5.109 5.110 5.111 US279 US340		
2190.5–2194 MARITIME MOBILE			2190.5–2194 MARITIME MOBILE (telephony) US340	2190.5–2194 MARITIME MOBILE US340	Maritime (80)
2194–2300 FIXED MOBILE except aeronautical mobile (R)	2194–2300 FIXED MOBILE		2194–2495 FIXED MOBILE	2194–2495 FIXED MOBILE except aeronautical mobile	Maritime (80) Private Land Mobile (90)
5.92 5.103 5.112	5.112		US340	US340	
2300–2498 FIXED MOBILE except aeronautical mobile (R) BROADCASTING 5.113	2300–2495 FIXED MOBILE BROADCASTING 5.113		US22 US340	US22 US340 NG7	
5.103	2495–2501 STANDARD FREQUENCY AND TIME SIGNAL (2500 kHz)		2495–2505 STANDARD FREQUENCY AND TIME SIGNAL (2500 kHz)		

(Continued)

TABLE F.1 (Continued)

Worldwide Frequency Allocations

Table of Frequency Allocations				1,800–3,025 kHz (MF/HF)		
International Table			United States Table			FCC Rule Part(s)
Region 1 Table	Region 2 Table	Region 3 Table	Federal Table	Non-Federal Table		
2498–2501 STANDARD FREQUENCY AND TIME SIGNAL (2500 kHz)			US1 US340			
2501–2502 STANDARD FREQUENCY AND TIME SIGNAL Space research						
2502–2625 FIXED MOBILE except aeronautical mobile (R) 5.92 5.103 5.114	2502–2505 STANDARD FREQUENCY AND TIME SIGNAL	2502–2505 STANDARD FREQUENCY AND TIME SIGNAL	2505–2850 FIXED MOBILE US285	2505–2850 FIXED MOBILE except aeronautical mobile US285		Maritime (80) Aviation (87) Private Land Mobile (90)
	2505–2850 FIXED MOBILE					
2625–2650 MARITIME MOBILE MARITIME RADIONAVIGATION 5.92						

(Continued)

TABLE F.1 (*Continued*)

Worldwide Frequency Allocations

Table of Frequency Allocations   1,800–3,025 kHz (MF/HF)

International Table			United States Table		FCC Rule Part(s)
Region 1 Table	Region 2 Table	Region 3 Table	Federal Table	Non-Federal Table	
2650–2850 FIXED MOBILE except aeronautical mobile (R) 5.92 5.103			US22 US340	US22 US340	
2850–3025 AERONAUTICAL MOBILE (R) 5.111 5.115			2850–3025 AERONAUTICAL MOBILE (R) 5.111 5.115 US283 US340	AERONAUTICAL MOBILE (R)	Aviation (87)

Table of Frequency Allocations   3.025–5.68 MHz (HF)

Region 1 Table	Region 2 Table	Region 3 Table	Federal Table	Non-Federal Table	FCC Rule Part(s)
3.025–3.155 AERONAUTICAL MOBILE (OR)			3.025–3.155 AERONAUTICAL MOBILE (OR) US340	AERONAUTICAL MOBILE (OR)	
3.155–3.2 FIXED MOBILE except aeronautical mobile (R) 5.116 5.117			3.155–3.23 FIXED MOBILE except aeronautical mobile (R)		Maritime (80) Private Land Mobile (90)

(*Continued*)

TABLE F.1 (Continued)
Worldwide Frequency Allocations

Table of Frequency Allocations   3.025–5.68 MHz (HF)

International Table			United States Table		FCC Rule Part(s)
Region 1 Table	**Region 2 Table**	**Region 3 Table**	**Federal Table**	**Non-Federal Table**	
3.2–3.23 FIXED MOBILE except aeronautical mobile (R) BROADCASTING 5.113	MOBILE except aeronautical mobile (R) BROADCASTING 5.113		US22 US340		
5.116 5.118	5.116				
3.23–3.4 FIXED MOBILE except aeronautical mobile BROADCASTING 5.113	3.23–3.4 FIXED MOBILE except aeronautical mobile BROADCASTING 5.113		3.23–3.4 FIXED MOBILE except aeronautical mobile Radiolocation	MOBILE except aeronautical mobile	Maritime (80) Aviation (87) Private Land Mobile (90)
5.116 5.118	5.116 5.118		US340		
3.4–3.5 AERONAUTICAL MOBILE (R)			3.4–3.5 AERONAUTICAL MOBILE (R)		Aviation (87)
			US283 US340		

(Continued)

TABLE F.1 (*Continued*)

Worldwide Frequency Allocations

Table of Frequency Allocations   3.025–5.68 MHz (HF)

International Table			United States Table		FCC Rule Part(s)
Region 1 Table	Region 2 Table	Region 3 Table	Federal Table	Non-Federal Table	
3.5–3.8 AMATEUR FIXED MOBILE except aeronautical mobile 5.92	3.5–3.75 AMATEUR 5.119	3.5–3.9 AMATEUR FIXED MOBILE	3.5–4	3.5–4 AMATEUR	Amateur Radio (97)
3.8–3.9 FIXED AERONAUTICAL MOBILE (OR) LAND MOBILE 3.9–3.95 AERONAUTICAL MOBILE (OR) 5.123	3.75–4 AMATEUR FIXED MOBILE except aeronautical mobile (R)	3.9–3.95 AERONAUTICAL MOBILE BROADCASTING			
3.95–4 FIXED BROADCASTING 5.122 5.125	5.122 5.125	3.95–4 FIXED BROADCASTING 5.126	US340	US340	

(*Continued*)

TABLE F.1 (*Continued*)
Worldwide Frequency Allocations

Table of Frequency Allocations — 3.025–5.68 MHz (HF)

International Table			United States Table		FCC Rule Part(s)
Region 1 Table	**Region 2 Table**	**Region 3 Table**	**Federal Table**	**Non-Federal Table**	
4–4.063 FIXED MARITIME MOBILE 5.127			4–4.063 FIXED MARITIME MOBILE		Maritime (80)
5.126			US340		
4.063–4.438 MARITIME MOBILE 5.79A 5.109 5.110 5.130 5.131 5.132			4.063–4.438 MARITIME MOBILE 5.79A 5.109 5.110 5.130 5.131 5.132 US82		Maritime (80) Aviation (87)
5.128			US296 US340		
4.438–4.65 FIXED MOBILE except aeronautical mobile (R)		4.438–4.65 FIXED MOBILE except aeronautical mobile	4.438–4.65 FIXED MOBILE except aeronautical mobile (R) US22 US340		Maritime (80) Aviation (87) Private Land Mobile (90)
4.65–4.7 AERONAUTICAL MOBILE (R)			4.65–4.7 AERONAUTICAL MOBILE (R) US282 US283 US340	AERONAUTICAL MOBILE (R)	Aviation (87)
4.7–4.75 AERONAUTICAL MOBILE (OR)			4.7–4.75 AERONAUTICAL MOBILE (OR) US340	AERONAUTICAL MOBILE (OR)	

(*Continued*)

TABLE F.1 (*Continued*)

Worldwide Frequency Allocations

Table of Frequency Allocations   **3.025–5.68 MHz (HF)**

International Table			United States Table		FCC Rule Part(s)
Region 1 Table	Region 2 Table	Region 3 Table	Federal Table	Non-Federal Table	
4.75–4.85 FIXED AERONAUTICAL MOBILE (OR) LAND MOBILE BROADCASTING 5.113	4.75–4.85 FIXED MOBILE except aeronautical mobile (R) BROADCASTING 5.113	4.75–4.85 FIXED BROADCASTING 5.113 Land mobile	4.75–4.85 FIXED MOBILE except aeronautical mobile (R) US340	MOBILE except aeronautical mobile (R) US340	Maritime (80) Private Land Mobile (90)
4.85–4.995 FIXED LAND MOBILE BROADCASTING 5.113	STANDARD FREQUENCY AND TIME SIGNAL (5 MHz)		4.85–4.995 FIXED MOBILE US340	4.85–4.995 FIXED US340	Aviation (87) Private Land Mobile (90)
4.995–5.003 STANDARD FREQUENCY AND TIME SIGNAL (5 MHz)			4.995–5.005 STANDARD FREQUENCY AND TIME SIGNAL (5 MHz) US1 US340		
5.003–5.005 STANDARD FREQUENCY AND TIME SIGNAL Space research					
5.005–5.06 FIXED BROADCASTING 5.113			5.005–5.06 FIXED US22 US340		Aviation (87) Private Land Mobile (90)

(*Continued*)

TABLE F.1 (*Continued*)

Worldwide Frequency Allocations

Table of Frequency Allocations

3.025–5.68 MHz (HF)

International Table			United States Table		
Region 1 Table	**Region 2 Table**	**Region 3 Table**	**Federal Table**	**Non-Federal Table**	**FCC Rule Part(s)**
5.06–5.25 FIXED Mobile except aeronautical mobile			5.06–5.45 FIXED US22 Mobile except aeronautical mobile		Maritime (80) Aviation (87) Private Land Mobile (90) Amateur Radio (97)
5.133					
5.25–5.45 FIXED MOBILE except aeronautical mobile			US23 US212 US340		
5.45–5.48 FIXED AERONAUTICAL MOBILE (OR) LAND MOBILE	5.45–5.48 AERONAUTICAL MOBILE (R)	5.45–5.48 FIXED AERONAUTICAL MOBILE (OR) LAND MOBILE	5.45–5.68 AERONAUTICAL MOBILE (R)	AERONAUTICAL MOBILE (R)	Aviation (87)
5.48–5.68 AERONAUTICAL MOBILE (R)					
5.111 5.115			5.111 5.115 US283 US340		

5.68–10.005 MHz (HF)

5.68–5.73 AERONAUTICAL MOBILE (OR)			5.68–5.73 AERONAUTICAL MOBILE (OR)	AERONAUTICAL MOBILE (OR)	
5.111 5.115			5.111 5.115 US340		

(Continued)

TABLE F.1 (Continued)

Worldwide Frequency Allocations

Table of Frequency Allocations   5.68–10.005 MHz (HF)

International Table			United States Table		FCC Rule Part(s)
Region 1 Table	Region 2 Table	Region 3 Table	Federal Table	Non-Federal Table	
5.73–5.9 FIXED LAND MOBILE	5.73–5.9 FIXED MOBILE except aeronautical mobile (R)	5.73–5.9 FIXED Mobile except aeronautical mobile (R)	5.73–5.9 FIXED MOBILE except aeronautical mobile (R) US340	MOBILE except aeronautical mobile (R)	Maritime (80) Aviation (87) Private Land Mobile (90)
5.9–5.95 BROADCASTING 5.134 5.136 5.95–6.2 BROADCASTING 5.137			5.9–6.2 BROADCASTING 5.134 US136 US340	BROADCASTING 5.134	International Broadcast Stations (73F)
6.2–6.525 MARITIME MOBILE 5.109 5.110 5.130 5.132 5.137	6.2–6.525 MARITIME MOBILE 5.109 5.110 5.130 5.132		6.2–6.525 MARITIME MOBILE 5.109 5.110 5.130 5.132 US82 US296 US340		Maritime (80)
6.525–6.685 AERONAUTICAL MOBILE (R)	6.525–6.685 AERONAUTICAL MOBILE (R)		6.525–6.685 AERONAUTICAL MOBILE (R) US283 US340		Aviation (87)
6.685–6.765 AERONAUTICAL MOBILE (OR)	6.685–6.765 AERONAUTICAL MOBILE (OR)		6.685–6.765 AERONAUTICAL MOBILE (OR) US340		

(Continued)

TABLE F.1 (Continued)

Worldwide Frequency Allocations

Table of Frequency Allocations   **5.68–10.005 MHz (HF)**

International Table			United States Table		FCC Rule Part(s)
Region 1 Table	**Region 2 Table**	**Region 3 Table**	**Federal Table**	**Non-Federal Table**	
6.765–7 FIXED MOBILE except aeronautical mobile (R) 5.138			6.765–7 FIXED US22 MOBILE except aeronautical mobile (R) 5.138 US340		ISM Equipment (18) Private Land Mobile (90)
7–7.1 AMATEUR AMATEUR-SATELLITE 5.140 5.141 5.141A			7–7.2	7–7.1 AMATEUR AMATEUR-SATELLITE US340	Amateur Radio (97)
7.1–7.2 AMATEUR 5.142 5.141A 5.141B			US340	7.1–7.2 AMATEUR US340	
7.2–7.3 BROADCASTING	7.2–7.3 AMATEUR 5.142	7.2–7.3 BROADCASTING	7.2–7.3 US142 US340	7.2–7.3 AMATEUR US142 US340	International Broadcast Stations (73F) Amateur Radio (97)
7.3–7.4 BROADCASTING 5.134 5.143 5.143A 5.143B 5.143C 5.143D			7.3–7.4 BROADCASTING 5.134 US136 US340		International Broadcast Stations (73F) Maritime (80) Private Land Mobile (90) *(Continued)*

TABLE F.1 (Continued)

Worldwide Frequency Allocations

Table of Frequency Allocations			5.68–10.005 MHz (HF)		
International Table			United States Table		
Region 1 Table	Region 2 Table	Region 3 Table	Federal Table	Non-Federal Table	FCC Rule Part(s)
7.4–7.45 BROADCASTING	7.4–7.45 FIXED MOBILE except aeronautical mobile (R)	7.4–7.45 BROADCASTING	7.4–7.45 FIXED	MOBILE except aeronautical mobile (R)	
5.143B 5.143C		5.143A 5.143C	US142 US340		
7.45–8.1 FIXED MOBILE except aeronautical mobile (R)			7.45–8.1 FIXED US22 MOBILE except aeronautical mobile (R)	MOBILE except aeronautical mobile (R)	Maritime (80) Aviation (87) Private Land Mobile (90)
5.144			US340		
8.1–8.195 FIXED MARITIME MOBILE			8.1–8.195 FIXED MARITIME MOBILE		Maritime (80)
			US340		
8.195–8.815 MARITIME MOBILE 5.109 5.110 5.132 5.145			8.195–8.815 MARITIME MOBILE 5.109 5.110 5.132 5.145 US82		Maritime (80) Aviation (87)
5.111			5.111 US296 US340		
8.815–8.965 AERONAUTICAL MOBILE (R)			8.815–8.965 AERONAUTICAL MOBILE (R)	AERONAUTICAL MOBILE (R)	Aviation (87)
			US340		

(Continued)

TABLE F.1 (Continued)

Worldwide Frequency Allocations

Table of Frequency Allocations					
			5.68–10.005 MHz (HF)		
International Table			United States Table		
Region 1 Table	Region 2 Table	Region 3 Table	Federal Table	Non-Federal Table	FCC Rule Part(s)
8.965–9.04 AERONAUTICAL MOBILE (OR)			8.965–9.04 AERONAUTICAL MOBILE (OR) US340	AERONAUTICAL MOBILE (OR)	
9.04–9.4 FIXED			9.04–9.4 FIXED US340		Maritime (80) Private Land Mobile (90)
9.4–9.5 BROADCASTING 5.134 5.146 9.5–9.9 BROADCASTING 5.147	BROADCASTING 5.134		9.4–9.9 BROADCASTING 5.134 US136 US340		International Broadcast Stations (73F)
9.9–9.995 FIXED			9.9–9.995 FIXED US340		Private Land Mobile (90)

(Continued)

TABLE F.1 (Continued)

Worldwide Frequency Allocations

Table of Frequency Allocations — 5.68–10.005 MHz (HF)

International Table			United States Table		FCC Rule Part(s)
Region 1 Table	Region 2 Table	Region 3 Table	Federal Table	Non-Federal Table	
9.995–10.003 STANDARD FREQUENCY AND TIME SIGNAL (10 MHz)			9.995–10.005 STANDARD FREQUENCY AND TIME SIGNAL (10 MHz)		
5.111					
10.003–10.005 STANDARD FREQUENCY AND TIME SIGNAL Space research					
5.111			5.111 US1 US340		

Table of Frequency Allocations — 10.005–15.01 MHz (HF)

International Table			United States Table		FCC Rule Part(s)
Region 1 Table	Region 2 Table	Region 3 Table	Federal Table	Non-Federal Table	
10.005–10.1 AERONAUTICAL MOBILE (R)			10.005–10.1 AERONAUTICAL MOBILE (R)	AERONAUTICAL MOBILE (R)	Aviation (87)
5.111			5.111 US283 US340		
10.1–10.15 FIXED Amateur			10.1–10.15 US247 US340	10.1–10.15 AMATEUR US247 US340	Amateur Radio (97)
10.15–11.175 FIXED Mobile except aeronautical mobile (R)			10.15–11.175 FIXED Mobile except aeronautical mobile (R) US340	Mobile except aeronautical mobile (R)	Private Land Mobile (90)
11.175–11.275 AERONAUTICAL MOBILE (OR)			11.175–11.275 AERONAUTICAL MOBILE (OR) US340		

(Continued)

TABLE F.1 (Continued)

Worldwide Frequency Allocations

Table of Frequency Allocations			10.005–15.01 MHz (HF)		
International Table			United States Table		
Region 1 Table	Region 2 Table	Region 3 Table	Federal Table	Non-Federal Table	FCC Rule Part(s)
11.275–11.4 AERONAUTICAL MOBILE (R)			11.275–11.4 AERONAUTICAL MOBILE (R) US283 US340		Aviation (87)
11.4–11.6 FIXED			11.4–11.6 FIXED US340		Private Land Mobile (90)
11.6–11.65 BROADCASTING 5.134			11.6–12.1 BROADCASTING 5.134		International Broadcast Stations (73F)
5.146					
11.65–12.05 BROADCASTING					
5.147					
12.05–12.1 BROADCASTING 5.134			US136 US340		
5.146					
12.1–12.23 FIXED			12.1–12.23 FIXED US340		Private Land Mobile (90)

(Continued)

TABLE F.1 (Continued)
Worldwide Frequency Allocations

Table of Frequency Allocations			10.005–15.01 MHz (HF)		
International Table			United States Table		FCC Rule Part(s)
Region 1 Table	Region 2 Table	Region 3 Table	Federal Table	Non-Federal Table	
12.23–13.2 MARITIME MOBILE 5.109 5.110 5.132 5.145	MARITIME MOBILE 5.109 5.110 5.132 5.145		12.23–13.2 MARITIME MOBILE 5.109 5.110 5.132 5.145 US82 US296 US340	12.23–13.2 MARITIME MOBILE 5.109 5.110 5.132 5.145	Maritime (80)
13.2–13.26 AERONAUTICAL MOBILE (OR)			13.2–13.26 AERONAUTICAL MOBILE (OR) US340	AERONAUTICAL MOBILE (OR)	
13.26–13.36 AERONAUTICAL MOBILE (R)			13.26–13.36 AERONAUTICAL MOBILE (R) US283 US340	AERONAUTICAL MOBILE (R)	Aviation (87)
13.36–13.41 FIXED RADIO ASTRONOMY 5.149			13.36–13.41 RADIO ASTRONOMY US342 G115	13.36–13.41 RADIO ASTRONOMY US342	
13.41–13.57 FIXED Mobile except aeronautical mobile (R) 5.150			13.41–13.57 FIXED Mobile except aeronautical mobile (R) 5.150 US340	13.41–13.57 FIXED 5.150 US340	ISM Equipment (18) Private Land Mobile (90)

(Continued)

TABLE F.1 (Continued)

Worldwide Frequency Allocations

Table of Frequency Allocations					10.005–15.01 MHz (HF)
International Table			United States Table		FCC Rule Part(s)
Region 1 Table	Region 2 Table	Region 3 Table	Federal Table	Non-Federal Table	
13.57—13.6 BROADCASTING 5.134 5.151			13.57—13.87 BROADCASTING 5.134		International Broadcast Stations (73F)
13.6—13.8 BROADCASTING					
13.8—13.87 BROADCASTING 5.134 5.151			US136 US340		
13.87—14 FIXED Mobile except aeronautical mobile (R)			13.87—14 FIXED Mobile except aeronautical mobile (R) US340	13.87—14 FIXED US340	Private Land Mobile (90)
14—14.25 AMATEUR AMATEUR-SATELLITE			14—14.35	14—14.25 AMATEUR AMATEUR-SATELLITE US340	Amateur Radio (97)
14.25—14.35 AMATEUR 5.152			US340	14.25—14.35 AMATEUR US340	

(Continued)

TABLE F.1 (Continued)

Worldwide Frequency Allocations

Table of Frequency Allocations

International Table			United States Table		FCC Rule Part(s)
Region 1 Table	Region 2 Table	Region 3 Table	Federal Table	Non-Federal Table	
10.005–15.01 MHz (HF)					
14.35–14.99 FIXED Mobile except aeronautical mobile (R)			14.35–14.99 FIXED Mobile except aeronautical mobile (R) US340	14.35–14.99 FIXED US340	Private Land Mobile (90)
14.99–15.005 STANDARD FREQUENCY AND TIME SIGNAL (15 MHz) 5.111			14.99–15.01 STANDARD FREQUENCY AND TIME SIGNAL (15 MHz)		
15.005–15.01 STANDARD FREQUENCY AND TIME SIGNAL Space research			5.111 US1 US340		
15.01–22.855 MHz (HF)					
15.01–15.1 AERONAUTICAL MOBILE (OR)			15.01–15.1 AERONAUTICAL MOBILE (OR) US340		
15.1–15.6 BROADCASTING			15.1–15.8 BROADCASTING 5.134	15.1–15.8 BROADCASTING 5.134	International Broadcast Stations (73F)
15.6–15.8 BROADCASTING 5.134 5.146			US136 US340		

(Continued)

TABLE F.1 (*Continued*)

Worldwide Frequency Allocations

Table of Frequency Allocations			15.01–22.855 MHz (HF)		
International Table			United States Table		
Region 1 Table	Region 2 Table	Region 3 Table	Federal Table	Non-Federal Table	FCC Rule Part(s)
15.8–16.36 FIXED 5.153			15.8–16.36 FIXED US340		Private Land Mobile (90)
16.36–17.41 MARITIME MOBILE 5.109 5.110 5.132 5.145			16.36–17.41 MARITIME MOBILE 5.109 5.110 5.132 5.145 US82 US296 US340		Maritime (80)
17.41–17.48 FIXED			17.41–17.48 FIXED US340		Private Land Mobile (90)
17.48–17.55 BROADCASTING 5.134 5.146			17.48–17.9 BROADCASTING 5.134		International Broadcast Stations (73F)
17.55–17.9 BROADCASTING			US136 US340		
17.9–17.97 AERONAUTICAL MOBILE (R)			17.9–17.97 AERONAUTICAL MOBILE (R) US283 US340		Aviation (87)

(Continued)

TABLE F.1 (*Continued*)

Worldwide Frequency Allocations

Table of Frequency Allocations			15.01–22.855 MHz (HF)		
International Table			United States Table		
Region 1 Table	Region 2 Table	Region 3 Table	Federal Table	Non-Federal Table	FCC Rule Part(s)
17.97–18.03 AERONAUTICAL MOBILE (OR)			17.97–18.03 AERONAUTICAL MOBILE (OR) US340	AERONAUTICAL MOBILE (OR)	
18.030–18.052 FIXED			18.03–18.068 FIXED		Maritime (80) Private Land Mobile (90)
18.052–18.068 FIXED Space research			US340		
18.068–18.168 AMATEUR AMATEUR-SATELLITE 5.154			18.068–18.168 US340	18.068–18.168 AMATEUR AMATEUR-SATELLITE US340	Amateur Radio (97)
18.168–18.78 FIXED Mobile except aeronautical mobile			18.168–18.78 FIXED Mobile US340		Maritime (80) Private Land Mobile (90)
18.78–18.9 MARITIME MOBILE			18.78–18.9 MARITIME MOBILE US82 US296 US340		Maritime (80)

(*Continued*)

TABLE F.1 (*Continued*)

Worldwide Frequency Allocations

Table of Frequency Allocations			15.01–22.855 MHz (HF)		
International Table			United States Table		
Region 1 Table	Region 2 Table	Region 3 Table	Federal Table	Non-Federal Table	FCC Rule Part(s)
18.9–19.02 BROADCASTING 5.134 5.146			18.9–19.02 BROADCASTING 5.134 US136 US340		International Broadcast Stations (73F)
19.02–19.68 FIXED			19.02–19.68 FIXED US340		Private Land Mobile (90)
19.68–19.8 MARITIME MOBILE 5.132			19.68–19.8 MARITIME MOBILE 5.132 US340		Maritime (80)
19.8–19.99 FIXED			19.8–19.99 FIXED US340		Private Land Mobile (90)
19.99–19.995 STANDARD FREQUENCY AND TIME SIGNAL Space research 5.111			19.99–20.01 STANDARD FREQUENCY AND TIME SIGNAL (20 MHz)		
19.995–20.01 STANDARD FREQUENCY AND TIME SIGNAL (20 MHz) 5.111			5.111 US1 US340		

(*Continued*)

TABLE F.1 (Continued)

Worldwide Frequency Allocations

Table of Frequency Allocations			15.01–22.855 MHz (HF)		
International Table			United States Table		FCC Rule Part(s)
Region 1 Table	Region 2 Table	Region 3 Table	Federal Table	Non-Federal Table	
20.01–21 FIXED Mobile			20.01–21 FIXED Mobile	20.01–21 FIXED	Private Land Mobile (90)
			US340	US340	
21–21.45 AMATEUR AMATEUR-SATELLITE			21–21.45	21–21.45 AMATEUR AMATEUR- SATELLITE	Amateur Radio (97)
			US340	US340	
21.45–21.85 BROADCASTING			21.45–21.85 BROADCASTING		International Broadcast Stations (73F)
			US340		
21.85–21.87 FIXED 5.155A 5.155			21.85–21.924 FIXED		Aviation (87) Private Land Mobile (90)
21.87–21.924 FIXED 5.155B			US340		
21.924–22 AERONAUTICAL MOBILE (R)			21.924–22 AERONAUTICAL MOBILE (R)		Aviation (87)
			US340		

(Continued)

TABLE F.1 (Continued)

Worldwide Frequency Allocations

Table of Frequency Allocations

15.01–22.855 MHz (HF)

International Table			United States Table		FCC Rule Part(s)
Region 1 Table	Region 2 Table	Region 3 Table	Federal Table	Non-Federal Table	
22–22.855 MARITIME MOBILE 5.132 5.156			22–22.855 MARITIME MOBILE 5.132 US296 US340	22–22.855 MARITIME MOBILE 5.132 US82	Maritime (80)

Table of Frequency Allocations

22.855–29.7 MHz (HF)

International Table			United States Table		FCC Rule Part(s)
Region 1 Table	Region 2 Table	Region 3 Table	Federal Table	Non-Federal Table	
22.855–23 FIXED 5.156			22.855–23 FIXED US340	22.855–23 FIXED	Private Land Mobile (90)
23–23.2 FIXED Mobile except aeronautical mobile (R) 5.156			23–23.2 FIXED Mobile except aeronautical mobile (R) US340	23–23.2 FIXED	
23.2–23.35 FIXED 5.156A AERONAUTICAL MOBILE (OR)			23.2–23.35 AERONAUTICAL MOBILE (OR) US340	US340	
23.35–24 FIXED MOBILE except aeronautical mobile 5.157			23.35–24.89 FIXED MOBILE except aeronautical mobile	23.35–24.89 FIXED	Private Land Mobile (90)
24–24.89 FIXED LAND MOBILE			US340	US340	

(*Continued*)

TABLE F.1 (Continued)

Worldwide Frequency Allocations

Table of Frequency Allocations			22.855–29.7 MHz (HF)		
International Table			United States Table		
Region 1 Table	Region 2 Table	Region 3 Table	Federal Table	Non-Federal Table	FCC Rule Part(s)
24.89–24.99 AMATEUR AMATEUR-SATELLITE			24.89–24.99 US340	24.89–24.99 AMATEUR AMATEUR-SATELLITE US340	Amateur Radio (97)
24.99–25.005 STANDARD FREQUENCY AND TIME SIGNAL (25 MHz)			24.99–25.01 STANDARD FREQUENCY AND TIME SIGNAL (25 MHz) US1 US340		
25.005–25.01 STANDARD FREQUENCY AND TIME SIGNAL Space research					
25.01–25.07 FIXED MOBILE except aeronautical mobile			25.01–25.07 US340	25.01–25.07 LAND MOBILE US340 NG112	Private Land Mobile (90)
25.07–25.21 MARITIME MOBILE			25.07–25.21 MARITIME MOBILE US82 US281 US296 US340	25.07–25.21 MARITIME MOBILE US82 US281 US296 US340 NG112	Maritime (80) Private Land Mobile (90)

(Continued)

TABLE F.1 (*Continued*)

Worldwide Frequency Allocations

Table of Frequency Allocations

22.855–29.7 MHz (HF)

International Table			United States Table		
Region 1 Table	Region 2 Table	Region 3 Table	Federal Table	Non-Federal Table	FCC Rule Part(s)
25.21–25.55 FIXED MOBILE except aeronautical mobile			25.21–25.33 US340	25.21–25.33 LAND MOBILE US340	Private Land Mobile (90)
			25.33–25.55 FIXED MOBILE except aeronautical mobile US340	25.33–25.55 US340	
25.55–25.67 RADIO ASTRONOMY 5.149			25.55–25.67 RADIO ASTRONOMY US74 US342	RADIO ASTRONOMY US74	
25.67–26.1 BROADCASTING			25.67–26.1 BROADCASTING US25 US340		International Broadcast Stations (73F) Remote Pickup (74D)
26.1–26.175 MARITIME MOBILE 5.132			26.1–26.175 MARITIME MOBILE 5.132 US25 US340	MARITIME MOBILE 5.132	Remote Pickup (74D) Low Power Auxiliary (74H) Maritime (80)

(*Continued*)

TABLE F.1 (*Continued*)
Worldwide Frequency Allocations

Table of Frequency Allocations　　22.855–29.7 MHz (HF)

International Table			United States Table		
Region 1 Table	Region 2 Table	Region 3 Table	Federal Table	Non-Federal Table	FCC Rule Part(s)
26.175–27.5 FIXED MOBILE except aeronautical mobile			26.175–26.48	26.175–26.48 LAND MOBILE	Remote Pickup (74D) Low Power Auxiliary (74H)
			US340	US340	
			26.48–26.95 FIXED MOBILE except aeronautical mobile	26.48–26.95	
			US340	US340	
			26.95–27.41	26.95–26.96 FIXED	ISM Equipment (18)
				5.150 US340	
				26.96–27.23 MOBILE except aeronautical mobile	ISM Equipment (18) Personal Radio (95)
				5.150 US340	
				27.23–27.41 FIXED MOBILE except aeronautical mobile	ISM Equipment (18) Private Land Mobile (90) Personal Radio (95)
5.150			5.150 US340	5.150 US340	

(*Continued*)

TABLE F.1 (*Continued*)

Worldwide Frequency Allocations

Table of Frequency Allocations | 22.855–29.7 MHz (HF)

International Table			United States Table		
Region 1 Table	Region 2 Table	Region 3 Table	Federal Table	Non-Federal Table	FCC Rule Part(s)
27.5–28 METEOROLOGICAL AIDS FIXED MOBILE			27.41–27.54 US340 27.54–28 FIXED MOBILE US298 US340	27.41–27.54 FIXED LAND MOBILE US340 27.54–28 US298 US340	Private Land Mobile (90)
28–29.7 AMATEUR AMATEUR-SATELLITE			28–29.7 US340	28–29.7 AMATEUR AMATEUR- SATELLITE US340	Amateur Radio (97)

(Continued)

TABLE F.1 (*Continued*)

Worldwide Frequency Allocations

Table of Frequency Allocations

29.7–47 MHz (HF/VHF)

International Table			United States Table		FCC Rule Part(s)
Region 1 Table	Region 2 Table	Region 3 Table	Federal Table	Non-Federal Table	
29.7–30.005 FIXED MOBILE			29.7–29.89	29.7–29.8 LAND MOBILE US340	Private Land Mobile (90)
				29.8–29.89 FIXED	
			US340	US340	
			29.89–29.91 FIXED MOBILE	29.89–29.91	
			US340	US340	
			29.91–30	29.91–30 FIXED	
			US340	US340	
			30–30.56 FIXED MOBILE	30–30.56	
30.005–30.01 SPACE OPERATION (satellite identification) FIXED MOBILE SPACE RESEARCH					

(*Continued*)

TABLE F.1 (*Continued*)

Worldwide Frequency Allocations

Table of Frequency Allocations

29.7–47 MHz (HF/VHF)

International Table			United States Table		FCC Rule Part(s)
Region 1 Table	Region 2 Table	Region 3 Table	Federal Table	Non-Federal Table	
30.01–37.5 FIXED MOBILE			30.56–32	30.56–32 FIXED LAND MOBILE NG124	Private Land Mobile (90)
			32–33 FIXED MOBILE	32–33	
			33–34	33–34 FIXED LAND MOBILE NG124	Private Land Mobile (90)
			34–35 FIXED MOBILE	34–35	
			35–36	35–36 FIXED LAND MOBILE	Public Mobile (22) Private Land Mobile (90)

(Continued)

TABLE F.1 (*Continued*)

Worldwide Frequency Allocations

Table of Frequency Allocations   29.7–47 MHz (HF/VHF)

International Table			United States Table		FCC Rule Part(s)
Region 1 Table	Region 2 Table	Region 3 Table	Federal Table	Non-Federal Table	
			36–37 FIXED MOBILE	36–37	
			US220	US220	
			37–37.5	37–37.5 LAND MOBILE NG124	Private Land Mobile (90)
37.5–38.25 FIXED MOBILE Radio astronomy			37.5–38 Radio astronomy	37.5–38 LAND MOBILE Radio astronomy	
			US342	US342 NG59 NG124	
			38–38.25 FIXED MOBILE RADIO ASTRONOMY	38–38.25 RADIO ASTRONOMY	
5.149			US81 US342	US81 US342	
38.25–39.986 FIXED MOBILE			38.25–39 FIXED MOBILE	38.25–39	

(Continued)

TABLE F.1 (*Continued*)

Worldwide Frequency Allocations

Table of Frequency Allocations 29.7–47 MHz (HF/VHF)

International Table			United States Table			FCC Rule Part(s)
Region 1 Table	Region 2 Table	Region 3 Table	Federal Table	Non-Federal Table		
39.986–40.02 FIXED MOBILE Space research			39–40	39–40 LAND MOBILE NG124		Private Land Mobile (90)
40.02–40.98 FIXED MOBILE			40–42 FIXED MOBILE	40–42		ISM Equipment (18) Private Land Mobile (90)
5.150						
40.98–41.015 FIXED MOBILE Space research						
5.160 5.161			5.150 US210 US220	5.150 US210 US220		
41.015–44 FIXED MOBILE			42–46.6	42–43.69 FIXED LAND MOBILE		Public Mobile (22) Private Land Mobile (90)
5.160 5.161				NG124 NG141		

(*Continued*)

TABLE F.1 (Continued)

Worldwide Frequency Allocations

Table of Frequency Allocations

29.7–47 MHz (HF/VHF)

International Table			United States Table		FCC Rule Part(s)
Region 1 Table	Region 2 Table	Region 3 Table	Federal Table	Non-Federal Table	
44-47 FIXED MOBILE				43.69-46.6 LAND MOBILE NG124 NG141	Private Land Mobile (90)
			46.6-47 FIXED MOBILE	46.6-47	
5.162 5.162A					

Table of Frequency Allocations

47–137 MHz (VHF)

International Table			United States Table		FCC Rule Part(s)
Region 1 Table	Region 2 Table	Region 3 Table	Federal Table	Non-Federal Table	
47-68 BROADCASTING	47-50 FIXED MOBILE	47-50 FIXED MOBILE BROADCASTING	47-49.6	47-49.6 LAND MOBILE NG124	Private Land Mobile (90)
			49.6-50 FIXED MOBILE	49.6-50	
	50-54 AMATEUR	5.162A	50-73	50-54 AMATEUR	Amateur Radio (97)
5.162A 5.166 5.167 5.167A 5.168 5.170					

(Continued)

TABLE F.1 (Continued)

Worldwide Frequency Allocations

Table of Frequency Allocations

	International Table			United States Table		
	Region 1 Table	Region 2 Table	Region 3 Table	Federal Table	Non-Federal Table	FCC Rule Part(s)
47–137 MHz (VHF)						
	5.162A 5.163 5.164 5.165 5.169 5.171	54-68 BROADCASTING Fixed Mobile 5.172	54-68 FIXED MOBILE BROADCASTING 5.162A		54-72 BROADCASTING	Broadcast Radio (TV)(73) LPTV, TV Translator/ Booster (74G) Low Power Auxiliary (74H)
	68–74.8 FIXED MOBILE except aeronautical mobile	68-72 BROADCASTING Fixed Mobile 5.173	68-74.8 FIXED MOBILE		NG5 NG14 NG115 NG149	
		72-73 FIXED MOBILE			72-73 FIXED MOBILE NG3 NG49 NG56	Public Mobile (22) Aviation (87) Private Land Mobile (90) Personal Radio (95)
		73-74.6 RADIO ASTRONOMY 5.178		73-74.6 RADIO ASTRONOMY US74 US246		

(Continued)

TABLE F.1 (Continued)

Worldwide Frequency Allocations

Table of Frequency Allocations

47–137 MHz (VHF)

International Table			United States Table		FCC Rule Part(s)
Region 1 Table	Region 2 Table	Region 3 Table	Federal Table	Non-Federal Table	
	74.6–74.8 FIXED MOBILE		74.6–74.8 FIXED MOBILE		Private Land Mobile (90)
5.149 5.175 5.177 5.179		5.149 5.176 5.179	US273		
74.8–75.2 AERONAUTICAL RADIONAVIGATION	74.8–75.2 AERONAUTICAL RADIONAVIGATION		74.8–75.2 AERONAUTICAL RADIONAVIGATION	AERONAUTICAL RADIONAVIGATION	Aviation (87)
5.180 5.181			5.180		
75.2–87.5 FIXED MOBILE except aeronautical mobile	75.2–75.4 FIXED MOBILE		75.2–75.4 FIXED MOBILE		Private Land Mobile (90)
	5.179		US273		
	75.4–76 FIXED MOBILE	75.4–87 FIXED MOBILE	75.4–88	75.4–76 FIXED MOBILE NG3 NG49 NG56	Public Mobile (22) Aviation (87) Private Land Mobile (90) Personal Radio (95)

(Continued)

TABLE F.1 (Continued)

Worldwide Frequency Allocations

Table of Frequency Allocations

47–137 MHz (VHF)

International Table			United States Table		FCC Rule Part(s)
Region 1 Table	Region 2 Table	Region 3 Table	Federal Table	Non-Federal Table	
5.175 5.179 5.187	76–88 BROADCASTING Fixed Mobile	5.182 5.183 5.188		76–88 BROADCASTING	Broadcast Radio (TV)(73) LPTV, TV Translator/Booster (74G) Low Power Auxiliary (74H)
87.5–100 BROADCASTING	5.185	87–100 FIXED MOBILE BROADCASTING		NG5 NG14 NG115 NG149	
5.190	88–100 BROADCASTING		88–108	88–108 BROADCASTING NG2	Broadcast Radio (FM) (73) FM Translator/Booster (74L)
100–108 BROADCASTING					
5.192 5.194			US93	US93 NG5	
108–117.975 AERONAUTICAL RADIONAVIGATION	108–117.975 AERONAUTICAL RADIONAVIGATION		108–117.975 AERONAUTICAL RADIONAVIGATION	AERONAUTICAL RADIONAVIGATION	Aviation (87)
5.197 5.197A			US93 US343		
117.975–137 AERONAUTICAL MOBILE (R)	AERONAUTICAL MOBILE (R)		117.975–121.9375 AERONAUTICAL MOBILE (R)		
			5.111 5.200 US26 US28 US36		

(Continued)

TABLE F.1 (Continued)

Worldwide Frequency Allocations

Table of Frequency Allocations   47–137 MHz (VHF)

International Table			United States Table		FCC Rule Part(s)
Region 1 Table	Region 2 Table	Region 3 Table	Federal Table	Non-Federal Table	
			121.9375–123.0875	121.9375–123.0875 AERONAUTICAL MOBILE	
			US30 US31 US33 US80 US102 US213	US30 US31 US33 US80 US102 US213	
			123.0875–123.5875 AERONAUTICAL MOBILE 5.200 US32 US33 US112		
			123.5875–128.8125 AERONAUTICAL MOBILE (R) US26 US36		
			128.8125–132.0125 AERONAUTICAL MOBILE (R)	128.8125–132.0125 AERONAUTICAL MOBILE (R)	
			132.0125–136 AERONAUTICAL MOBILE (R) US26		
			136–137 US244	136–137 AERONAUTICAL MOBILE (R) US244	

5.111 5.200 5.201 5.202

(*Continued*)

TABLE F.1 (*Continued*)

Worldwide Frequency Allocations

Table of Frequency Allocations 137–156.7625 MHz (VHF)

International Table			United States Table		FCC Rule Part(s)
Region 1 Table	**Region 2 Table**	**Region 3 Table**	**Federal Table**	**Non-Federal Table**	
137–137.025 SPACE OPERATION (space-to-Earth) METEOROLOGICAL-SATELLITE (space-to-Earth) MOBILE-SATELLITE (space-to-Earth) 5.208A 5.208B 5.209 SPACE RESEARCH (space-to-Earth) Fixed Mobile except aeronautical mobile (R)			137–137.025 SPACE OPERATION (space-to-Earth) METEOROLOGICAL-SATELLITE (space-to-Earth) MOBILE-SATELLITE (space-to-Earth) US319 US320 SPACE RESEARCH (space-to-Earth)	137–137.025 SPACE OPERATION (space-to-Earth) METEOROLOGICAL-SATELLITE (space-to-Earth) MOBILE-SATELLITE (space-to-Earth) US319 SPACE RESEARCH (space-to-Earth)	Satellite Communications (25)
5.204 5.205 5.206 5.207 5.208			5.208		
137.025–137.175 SPACE OPERATION (space-to-Earth) METEOROLOGICAL-SATELLITE (space-to-Earth) SPACE RESEARCH (space-to-Earth) Fixed Mobile-satellite (space-to-Earth) 5.208A 5.208B 5.209 Mobile except aeronautical mobile (R)			137.025–137.175 SPACE OPERATION (space-to-Earth) METEOROLOGICAL-SATELLITE (space-to-Earth) SPACE RESEARCH (space-to-Earth) Mobile-satellite (space-to-Earth) US319 US320		
5.204 5.205 5.206 5.207 5.208			5.208		

(*Continued*)

TABLE F.1 (*Continued*)

Worldwide Frequency Allocations

Table of Frequency Allocations | 137–156.7625 MHz (VHF)

International Table			United States Table		FCC Rule Part(s)
Region 1 Table	Region 2 Table	Region 3 Table	Federal Table	Non-Federal Table	
137.175–137.825 SPACE OPERATION (space-to-Earth) METEOROLOGICAL-SATELLITE (space-to-Earth) MOBILE-SATELLITE (space-to-Earth) 5.208A 5.208B 5.209 SPACE RESEARCH (space-to-Earth) Fixed Mobile except aeronautical mobile (R)			137.175–137.825 SPACE OPERATION (space-to-Earth) METEOROLOGICAL-SATELLITE (space-to-Earth) MOBILE-SATELLITE (space-to-Earth) US319 US320 SPACE RESEARCH (space-to-Earth)	SPACE OPERATION (space-to-Earth) METEOROLOGICAL-SATELLITE (space-to-Earth) MOBILE-SATELLITE (space-to-Earth) US319 SPACE RESEARCH (space-to-Earth)	
5.204 5.205 5.206 5.207 5.208			5.208		
137.825–138 SPACE OPERATION (space-to-Earth) METEOROLOGICAL-SATELLITE (space-to-Earth) SPACE RESEARCH (space-to-Earth) Fixed Mobile-satellite (space-to-Earth) 5.208A 5.208B 5.209 Mobile except aeronautical mobile (R)			137.825–138 SPACE OPERATION (space-to-Earth) METEOROLOGICAL-SATELLITE (space-to-Earth) SPACE RESEARCH (space-to-Earth) Mobile-satellite (space-to-Earth) US319 US320	SPACE OPERATION (space-to-Earth) METEOROLOGICAL-SATELLITE (space-to-Earth) SPACE RESEARCH (space-to-Earth) Mobile-satellite (space-to-Earth) US319 US320	
5.204 5.205 5.206 5.207 5.208			5.208		

(Continued)

TABLE F.1 (Continued)

Worldwide Frequency Allocations

Table of Frequency Allocations   137–156.7625 MHz (VHF)

International Table			United States Table		FCC Rule Part(s)
Region 1 Table	Region 2 Table	Region 3 Table	Federal Table	Non-Federal Table	
138–143.6 AERONAUTICAL MOBILE (OR)	138–143.6 FIXED MOBILE RADIOLOCATION Space research (space-to-Earth)	138–143.6 FIXED MOBILE Space research (space-to-Earth)	138–144 FIXED MOBILE	138–144	
5.210 5.211 5.212 5.214		5.207 5.213			
143.6–143.65 AERONAUTICAL MOBILE (OR) SPACE RESEARCH (space-to-Earth)	143.6–143.65 FIXED MOBILE RADIOLOCATION SPACE RESEARCH (space-to-Earth)	143.6–143.65 FIXED MOBILE SPACE RESEARCH (space-to-Earth)			
5.211 5.212 5.214		5.207 5.213			
143.65–144 AERONAUTICAL MOBILE (OR)	143.65–144 FIXED MOBILE RADIOLOCATION Space research (space-to-Earth)	143.65–144 FIXED MOBILE Space research (space-to-Earth)	G30		
5.210 5.211 5.212 5.214		5.207 5.213			

(Continued)

TABLE F.1 (Continued)

Worldwide Frequency Allocations

Table of Frequency Allocations   137–156.7625 MHz (VHF)

International Table			United States Table		FCC Rule Part(s)
Region 1 Table	Region 2 Table	Region 3 Table	Federal Table	Non-Federal Table	
144–146 AMATEUR AMATEUR-SATELLITE 5.216			144–148	144–146 AMATEUR AMATEUR-SATELLITE	Amateur Radio (97)
146–148 FIXED MOBILE except aeronautical mobile (R) 5.217	146–148 AMATEUR 5.217	146–148 AMATEUR FIXED MOBILE 5.217		146–148 AMATEUR	
148–149.9 FIXED MOBILE MOBILE-SATELLITE (Earth-to-space) 5.209	148–149.9 FIXED MOBILE MOBILE-SATELLITE (Earth-to-space) 5.209		148–149.9 FIXED MOBILE MOBILE-SATELLITE (Earth-to-space) US319 US320 US323 US325	148–149.9 MOBILE-SATELLITE (Earth-to-space) US320 US323 US325	Satellite Communications (25)
5.218 5.219 5.221	5.218 5.219 5.221		5.218 5.219 G30	5.218 5.219 US319	

(Continued)

TABLE F.1 (Continued)

Worldwide Frequency Allocations

Table of Frequency Allocations — 137–156.7625 MHz (VHF)

International Table			United States Table		FCC Rule Part(s)
Region 1 Table	Region 2 Table	Region 3 Table	Federal Table	Non-Federal Table	
149.9–150.05 MOBILE-SATELLITE (Earth-to-space) 5.209 5.224A RADIONAVIGATION-SATELLITE 5.224B			149.9–150.05 MOBILE-SATELLITE (Earth-to-space) US320 RADIONAVIGATION-SATELLITE	149.9–150.05 MOBILE-SATELLITE (Earth-to-space) US319 RADIONAVIGATION-SATELLITE	
5.220 5.222 5.223			5.223		
150.05–153 FIXED MOBILE except aeronautical mobile RADIO ASTRONOMY	150.05–156.4875 FIXED MOBILE		150.05–150.8 FIXED MOBILE	150.05–150.8	
			US73 G30	US73	
			150.8–152.855	150.8–152.855 FIXED LAND MOBILE NG4 NG51 NG112	Public Mobile (22) Private Land Mobile (90) Personal Radio (95)
5.149			US73	US73 NG124	
153–154 FIXED MOBILE except aeronautical mobile (R) Meteorological aids			152.855–156.2475	152.855–154 LAND MOBILE NG4	Remote Pickup (74D) Private Land Mobile (90)
				NG124	

(Continued)

TABLE F.1 (Continued)

Worldwide Frequency Allocations

Table of Frequency Allocations 137–156.7625 MHz (VHF)

International Table			United States Table		FCC Rule Part(s)
Region 1 Table	Region 2 Table	Region 3 Table	Federal Table	Non-Federal Table	
154–156.4875 FIXED MOBILE except aeronautical mobile (R) 5.226	5.225 5.226			154–156.2475 FIXED LAND MOBILE NG112 5.226 NG117 NG124 NG148	Maritime (80) Private Land Mobile (90) Personal Radio (95)
156.4875–156.5625 MARITIME MOBILE (distress and calling via DSC) 5.111 5.226 5.227			156.2475–156.7625 US77 US106 US226 US266	156.2475–156.7625 MARITIME MOBILE US106 US226 NG117	Maritime (80) Aviation (87)
156.5625–156.7625 FIXED MOBILE except aeronautical mobile (R) 5.226	156.5625–156.7625 FIXED MOBILE 5.225 5.226			US77 US266 NG124	

Table of Frequency Allocations 156.7625–267 MHz (VHF)

International Table			United States Table		FCC Rule Part(s)
156.7625–156.8375 MARITIME MOBILE (distress and calling) 5.111 5.226	156.7625–156.8375 MARITIME MOBILE (distress and calling)		156.7625–156.8375 MARITIME MOBILE (distress, urgency, safety and calling) 5.226 US266		Maritime (80) Aviation (87)

(Continued)

TABLE F.1 (Continued)

Worldwide Frequency Allocations

Table of Frequency Allocations   156.7625–267 MHz (VHF)

International Table			United States Table		FCC Rule Part(s)
Region 1 Table	Region 2 Table	Region 3 Table	Federal Table	Non-Federal Table	
156.8375–174 FIXED MOBILE except aeronautical mobile	156.8375–174 FIXED MOBILE		156.8375–157.0375	156.8375–157.0375 MARITIME MOBILE	
			5.226 US77 US266	5.226 US77 US266	
			157.0375–157.1875 MARITIME MOBILE US214	157.0375–157.1875	Maritime (80)
			5.226 US266 G109	5.226 US214 US266	
			157.1875–161.575	157.1875–157.45 MOBILE except aeronautical mobile US266	Maritime (80) Aviation (87) Private Land Mobile (90)
				5.226 NG111	
				157.45–161.575 FIXED LAND MOBILE NG28 NG111 NG112	Public Mobile (22) Remote Pickup (74D) Maritime (80) Private Land Mobile (90)
				5.226 NG6 NG70 NG124 NG148 NG155	

(Continued)

TABLE F.1 (Continued)
Worldwide Frequency Allocations

Table of Frequency Allocations			156.7625–267 MHz (VHF)		
International Table			United States Table		
Region 1 Table	Region 2 Table	Region 3 Table	Federal Table	Non-Federal Table	FCC Rule Part(s)
			161.575–161.625	161.575–161.625 MARITIME MOBILE	Public Mobile (22) Maritime (80)
			5.226 US77	5.226 US77 NG6 NG17	
			161.625–161.9625	161.625–161.775 LAND MOBILE NG6	Public Mobile (22) Remote Pickup (74D) Low Power Auxiliary (74H)
				5.226	
				161.775–161.9625 MOBILE except aeronautical mobile US266 NG6	Maritime (80) Private Land Mobile (90)
			US266	5.226	
			161.9625–161.9875 MARITIME MOBILE (AIS) US228		
			161.9875–162.0125	161.9875–162.0125 MOBILE except aeronautical mobile	Maritime (80)
				5.226	
			162.0125–162.0375 MARITIME MOBILE (AIS) US228		

(Continued)

TABLE F.1 (Continued)
Worldwide Frequency Allocations

Table of Frequency Allocations — 156.7625–267 MHz (VHF)

International Table			United States Table		FCC Rule Part(s)
Region 1 Table	Region 2 Table	Region 3 Table	Federal Table	Non-Federal Table	
5.226 5.227A 5.229			162.0375–173.2 FIXED MOBILE US8 US11 US13 US73 US300 US312 G5	162.0375–173.2 US8 US11 US13 US73 US300 US312	Remote Pickup (74D) Private Land Mobile (90)
			173.2–173.4	173.2–173.4 FIXED Land mobile	Private Land Mobile (90)
			173.4–174 FIXED MOBILE	173.4–174	
174–223 BROADCASTING	174–216 BROADCASTING Fixed Mobile 5.234	174–223 FIXED MOBILE BROADCASTING	G5 174–216	174–216 BROADCASTING NG5 NG14 NG115 NG149	Broadcast Radio (TV) (73) LPTV, TV Translator/ Booster (74G) Low Power Auxiliary (74H)

5.226 5.227A 5.230 5.231 5.232

(Continued)

TABLE F.1 (Continued)

Worldwide Frequency Allocations

Table of Frequency Allocations					156.7625–267 MHz (VHF)
International Table			United States Table		FCC Rule Part(s)
Region 1 Table	Region 2 Table	Region 3 Table	Federal Table	Non-Federal Table	
	216–220 FIXED MARITIME MOBILE Radiolocation 5.241		216–217 Fixed Land mobile	216–219 FIXED MOBILE except aeronautical mobile	Maritime (80) Private Land Mobile (90) Personal Radio (95)
			US210 US241 G2	US210 US241 NG173	
			217–220 Fixed Mobile	219–220 FIXED MOBILE except aeronautical mobile Amateur NG152	Maritime (80) Private Land Mobile (90) Amateur Radio (97)
	5.242		US210 US241	US210 US241 NG173	
	220–225 AMATEUR FIXED MOBILE Radiolocation 5.241		220–222 FIXED LAND MOBILE		Private Land Mobile (90)
5.235 5.237 5.243		5.233 5.238 5.240 5.245	US241 US242		

(Continued)

TABLE F.1 (Continued)

Worldwide Frequency Allocations

Table of Frequency Allocations — 156.7625–267 MHz (VHF)

	International Table		United States Table		
Region 1 Table	Region 2 Table	Region 3 Table	Federal Table	Non-Federal Table	FCC Rule Part(s)
223–230 BROADCASTING Fixed Mobile	225–235 FIXED MOBILE	223–230 FIXED MOBILE BROADCASTING AERONAUTICAL RADIO-NAVIGATION Radiolocation	222–225	222–225 AMATEUR	Amateur Radio (97)
		5.250	225–235 FIXED MOBILE	225–235	
5.243 5.246 5.247					
230–235 FIXED MOBILE		230–235 FIXED MOBILE AERONAUTICAL RADIO-NAVIGATION			
		5.250			
5.247 5.251 5.252			G27	235–267	
235–267 FIXED MOBILE			235–267 FIXED MOBILE		
5.111 5.252 5.254 5.256 5.256A			5.111 5.256 G27 G100	5.111 5.256	

(Continued)

TABLE F.1 (*Continued*)

Worldwide Frequency Allocations

Table of Frequency Allocations

267–410 MHz (VHF/UHF)

International Table			United States Table		FCC Rule Part(s)
Region 1 Table	Region 2 Table	Region 3 Table	Federal Table	Non-Federal Table	
267–272 FIXED MOBILE Space operation (space-to-Earth)			267–322 FIXED MOBILE	267–322	
5.254 5.257					
272–273 SPACE OPERATION (space-to-Earth) FIXED MOBILE					
5.254					
273–312 FIXED MOBILE					
5.254					
312–315 FIXED MOBILE Mobile-satellite (Earth-to-space) 5.254 5.255					

(Continued)

TABLE F.1 (*Continued*)

Worldwide Frequency Allocations

Table of Frequency Allocations | 267–410 MHz (VHF/UHF)

International Table			United States Table		FCC Rule Part(s)
Region 1 Table	Region 2 Table	Region 3 Table	Federal Table	Non-Federal Table	
315–322 FIXED MOBILE 5.254			G27 G100		
322–328.6 FIXED MOBILE RADIO ASTRONOMY 5.149			322–328.6 FIXED MOBILE US342 G27	322–328.6 US342	
328.6–335.4 AERONAUTICAL RADIONAVIGATION 5.258 5.259			328.6–335.4 AERONAUTICAL RADIONAVIGATION 5.258		Aviation (87)
335.4–387 FIXED MOBILE 5.254			335.4–399.9 FIXED MOBILE	335.4–399.9	
387–390 FIXED MOBILE Mobile-satellite (space-to-Earth) 5.208A 5.208B 5.254 5.255					

(*Continued*)

TABLE F.1 (Continued)
Worldwide Frequency Allocations

Table of Frequency Allocations				267–410 MHz (VHF/UHF)		
International Table			United States Table			
Region 1 Table	Region 2 Table	Region 3 Table	Federal Table	Non-Federal Table		FCC Rule Part(s)
390–399.9 FIXED MOBILE						
5.254			G27 G100			
399.9–400.05 MOBILE-SATELLITE (Earth-to-space) 5.209 5.224A RADIONAVIGATION-SATELLITE 5.222 5.224B 5.260			399.9–400.05 MOBILE-SATELLITE (Earth-to-space) US320 RADIONAVIGATION-SATELLITE 5.260	399.9–400.05 MOBILE-SATELLITE (Earth-to-space) US319 RADIONAVIGATION-SATELLITE 5.260		Satellite Communications (25)
5.220						
400.05–400.15 STANDARD FREQUENCY AND TIME SIGNAL-SATELLITE (400.1 MHz)			400.05–400.15 STANDARD FREQUENCY AND TIME SIGNAL-SATELLITE (400.1 MHz)	400.05–400.15 STANDARD FREQUENCY AND TIME SIGNAL-SATELLITE (400.1 MHz)		
5.261 5.262			5.261			

(Continued)

TABLE F.1 (*Continued*)

Worldwide Frequency Allocations

Table of Frequency Allocations			267–410 MHz (VHF/UHF)		
International Table			United States Table		FCC Rule Part(s)
Region 1 Table	Region 2 Table	Region 3 Table	Federal Table	Non-Federal Table	
400.15–401 METEOROLOGICAL AIDS METEOROLOGICAL-SATELLITE (space-to-Earth) MOBILE-SATELLITE (space-to-Earth) 5.208A 5.208B 5.209 SPACE RESEARCH (space-to-Earth) 5.263 Space operation (space-to-Earth)			400.15–401 METEOROLOGICAL AIDS (radiosonde) US70 METEORO-LOGICAL-SATELLITE (space-to-Earth) MOBILE-SATELLITE (space-to-Earth) US319 US320 US324 SPACE RESEARCH (space-to-Earth) 5.263 Space operation (space-to-Earth)	400.15–401 METEOROLOGICAL AIDS (radiosonde) US70 MOBILE-SATELLITE (space-to-Earth) US320 US324 SPACE RESEARCH (space-to-Earth) 5.263 Space operation (space-to-Earth)	Satellite Communications (25)
5.262 5.264			5.264	5.264 US319	

(*Continued*)

TABLE F.1 (*Continued*)

Worldwide Frequency Allocations

Table of Frequency Allocations

267–410 MHz (VHF/UHF)

International Table			United States Table		FCC Rule Part(s)
Region 1 Table	Region 2 Table	Region 3 Table	Federal Table	Non-Federal Table	
401–402 METEOROLOGICAL AIDS SPACE OPERATION (space-to-Earth) EARTH EXPLORATION-SATELLITE (Earth-to-space) METEOROLOGICAL-SATELLITE (Earth-to-space) Fixed Mobile except aeronautical mobile			401–402 METEOROLOGICAL AIDS (radiosonde) US70 SPACE OPERATION (space-to-Earth) EARTH EXPLORATION- SATELLITE (Earth-to-space) METEORO- LOGICAL- SATELLITE (Earth-to-space) US64 US384	401–402 METEOROLOGICAL AIDS (radiosonde) US70 SPACE OPERATION (space-to-Earth) Earth exploration-satellite (Earth-to-space) Meteorological- satellite (Earth-to-space) US64 US384	MedRadio (95I)

(Continued)

TABLE F.1 (*Continued*)
Worldwide Frequency Allocations

| Table of Frequency Allocations | | | 267–410 MHz (VHF/UHF) | | |
| International Table | | | United States Table | | |
Region 1 Table	Region 2 Table	Region 3 Table	Federal Table	Non-Federal Table	FCC Rule Part(s)
402–403 METEOROLOGICAL AIDS EARTH EXPLORATION-SATELLITE (Earth-to-space) METEOROLOGICAL-SATELLITE (Earth-to-space) Fixed Mobile except aeronautical mobile			402–403 METEOROLOGICAL AIDS (radiosonde) US70 EARTH EXPLORATION-SATELLITE (Earth-to-space) METEOROLOGICAL-SATELLITE (Earth-to-space) US64 US384	402–403 METEOROLOGICAL AIDS (radiosonde) US70 Earth exploration-satellite (Earth-to-space) Meteorological-satellite (Earth-to-space) US64 US384	
403–406 METEOROLOGICAL AIDS Fixed Mobile except aeronautical mobile			403–406 METEOROLOGICAL AIDS (radiosonde) US70 US64 G6	403–406 METEOROLOGICAL AIDS (radiosonde) US70 US64	
406–406.1 MOBILE-SATELLITE (Earth-to-space) 5.266 5.267			406–406.1 MOBILE-SATELLITE (Earth-to-space) 5.266 5.267		Maritime (EPIRBs) (80V) Aviation (ELTs) (87F) Personal Radio (95)

(*Continued*)

TABLE F.1 (*Continued*)

Worldwide Frequency Allocations

Table of Frequency Allocations   267–410 MHz (VHF/UHF)

International Table			United States Table		FCC Rule Part(s)
Region 1 Table	Region 2 Table	Region 3 Table	Federal Table	Non-Federal Table	
406.1–410 FIXED MOBILE except aeronautical mobile RADIO ASTRONOMY			406.1–410 FIXED MOBILE RADIO ASTRONOMY US74	406.1–410 RADIO ASTRONOMY US74	Private Land Mobile (90)
5.149			US13 US117 G5 G6	US13 US117	

Table of Frequency Allocations   410–698 MHz (UHF)

Region 1 Table	Region 2 Table	Region 3 Table	Federal Table	Non-Federal Table	FCC Rule Part(s)
410–420 FIXED MOBILE except aeronautical mobile SPACE RESEARCH (space-to-space) 5.268			410–420 FIXED MOBILE SPACE RESEARCH (space-to-space) 5.268	410–420	Private Land Mobile (90) MedRadio (95I)
			US13 US64 G5	US13 US64	
420–430 FIXED MOBILE except aeronautical mobile Radiolocation 5.269 5.270 5.271			420–450 RADIOLOCATION G2 G129	420–450 Amateur US270	Private Land Mobile (90) MedRadio (95I) Amateur Radio (97)

(*Continued*)

TABLE F.1 (*Continued*)

Worldwide Frequency Allocations

Table of Frequency Allocations — 410–698 MHz (UHF)

	International Table			United States Table		FCC Rule Part(s)
Region 1 Table	Region 2 Table	Region 3 Table		Federal Table	Non-Federal Table	
430–432 AMATEUR RADIOLOCATION 5.271 5.272 5.273 5.274 5.275 5.276 5.277	430–432 RADIOLOCATION Amateur 5.271 5.276 5.277 5.278 5.279					
432–438 AMATEUR RADIOLOCATION Earth exploration- satellite (active) 5.279A 5.138 5.271 5.272 5.276 5.277 5.280 5.281 5.282	432–438 RADIOLOCATION Amateur Earth exploration-satellite (active) 5.279A 5.271 5.276 5.277 5.278 5.279 5.281 5.282	Earth exploration-satellite (active) 5.279A				
438–440 AMATEUR RADIOLOCATION 5.271 5.273 5.274 5.275 5.276 5.277 5.283	438–440 RADIOLOCATION Amateur 5.271 5.276 5.277 5.278 5.279					

(Continued)

TABLE F.1 (*Continued*)

Worldwide Frequency Allocations

Table of Frequency Allocations

410–698 MHz (UHF)

International Table			United States Table		FCC Rule Part(s)
Region 1 Table	Region 2 Table	Region 3 Table	Federal Table	Non-Federal Table	
440–450 FIXED MOBILE except aeronautical mobile Radiolocation			5.286 US64 US87 US230 US269 US270 US397 G8	5.282 5.286 US64 US87 US230 US269 US397	
5.269 5.270 5.271 5.284 5.285 5.286					
450–455 FIXED MOBILE 5.286AA			450–454	450–454 LAND MOBILE	Remote Pickup (74D) Low Power Auxiliary (74H) Private Land Mobile (90) MedRadio (95I)
			5.286 US64 US87	5.286 US64 US87 NG112 NG124	
			454–456	454–455 FIXED LAND MOBILE	Public Mobile (22) Maritime (80) MedRadio (95I)
5.209 5.271 5.286 5.286A 5.286B 5.286C 5.286D 5.286E				US64 NG32 NG112 NG148	

(Continued)

TABLE F.1 (Continued)

Worldwide Frequency Allocations

Table of Frequency Allocations — 410–698 MHz (UHF)

International Table			United States Table		FCC Rule Part(s)
Region 1 Table	Region 2 Table	Region 3 Table	Federal Table	Non-Federal Table	
455–456 FIXED MOBILE 5.286AA	455–456 FIXED MOBILE 5.286AA MOBILE-SATELLITE (Earth-to-space) 5.286A 5.286B	455–456 FIXED MOBILE 5.286AA		455–456 LAND MOBILE	Remote Pickup (74D) Low Power Auxiliary (74H) MedRadio (95I)
5.209 5.271 5.286A 5.286B 5.286C 5.286E	5.286C	5.209 5.271 5.286A 5.286B 5.286C 5.286E	US64	US64	
456–459 FIXED MOBILE 5.286AA	5.209		456–459	456–460 FIXED LAND MOBILE	Public Mobile (22) Maritime (80) Private Land Mobile (90) MedRadio (95I)
5.271 5.287 5.288			5.287 US64 US288		
459–460 FIXED MOBILE 5.286AA MOBILE-SATELLITE (Earth-to-space) 5.286A 5.286B 5.286C	459–460 FIXED MOBILE 5.286AA MOBILE-SATELLITE (Earth-to-space) 5.286A 5.286B 5.286C	459–460 FIXED MOBILE 5.286AA	459–460		
5.209 5.271 5.286A 5.286B 5.286C 5.286E	5.209	5.209 5.271 5.286A 5.286B 5.286C 5.286E		5.287 US64 US288 NG32 NG112 NG124 NG148	

(Continued)

TABLE F.1 (*Continued*)

Worldwide Frequency Allocations

Table of Frequency Allocations

410–698 MHz (UHF)

International Table			United States Table		FCC Rule Part(s)
Region 1 Table	Region 2 Table	Region 3 Table	Federal Table	Non-Federal Table	
460–470 FIXED MOBILE 5.286AA Meteorological-satellite (space-to-Earth)			460–470 Meteorological-satellite (space-to-Earth)	460–462.5375 FIXED LAND MOBILE US209 US289 NG124	Private Land Mobile (90)
				462.5375–462.7375 LAND MOBILE US289	Personal Radio (95)
				462.7375–467.5375 FIXED LAND MOBILE 5.287 US73 US209 US288 US289 NG124	Maritime (80) Private Land Mobile (90)
				467.5375–467.7375 LAND MOBILE 5.287 US288 US289	Maritime (80) Personal Radio (95)

(Continued)

TABLE F.1 (*Continued*)

Worldwide Frequency Allocations

Table of Frequency Allocations

410–698 MHz (UHF)

International Table			United States Table		FCC Rule Part(s)
Region 1 Table	Region 2 Table	Region 3 Table	Federal Table	Non-Federal Table	
5.287 5.288 5.289 5.290				467.7375–470 FIXED LAND MOBILE	Maritime (80) Private Land Mobile (90)
470–790 BROADCASTING	470–512 BROADCASTING Fixed Mobile 5.292 5.293	470–585 FIXED MOBILE BROADCASTING	5.287 US73 US209 US288 US289	US73 US288 US289 NG124	
			470–608	470–512 FIXED LAND MOBILE BROADCASTING NG5 NG14 NG66 NG115 NG149	Public Mobile (22) Broadcast Radio (TV)(73) LPTV, TV Translator/ Booster (74G) Low Power Auxiliary (74H) Private Land Mobile (90)
		5.291 5.298 585–610 FIXED MOBILE BROADCASTING RADIO- NAVIGATION			
	512–608 BROADCASTING			512–608 FIXED MOBILE BROADCASTING NG5 NG14 NG115 NG149	Wireless Communications (27) Broadcast Radio (TV)(73) LPTV, TV Translator/ Booster (74G) Low Power Auxiliary (74H)
	5.297	5.149 5.305 5.306 5.307			

(Continued)

TABLE F.1 (Continued)

Worldwide Frequency Allocations

Table of Frequency Allocations · 410–698 MHz (UHF)

International Table			United States Table		FCC Rule Part(s)
Region 1 Table	Region 2 Table	Region 3 Table	Federal Table	Non-Federal Table	
	608–614 RADIO ASTRONOMY Mobile-satellite except aeronautical mobile-satellite (Earth-to-space)	610–890 FIXED MOBILE 5.313A 5.317A BROADCASTING	608–614 LAND MOBILE (medical telemetry and medical telecommand) RADIO ASTRONOMY US74 US246		Personal Radio (95)
5.149 5.291A 5.294 5.296 5.300 5.302 5.304 5.306 5.311A 5.312	614–698 BROADCASTING Fixed Mobile 5.293 5.309 5.311A	5.149 5.305 5.306 5.307 5.311A 5.320	614–698	614–698 FIXED MOBILE BROADCASTING NG5 NG14 NG115 NG149	Wireless Communications (27) Broadcast Radio (TV) (73) LPTV, TV Translator/Booster (74G) Low Power Auxiliary (74H)

(Continued)

TABLE F.1 (*Continued*)

Worldwide Frequency Allocations

Table of Frequency Allocations — 698–941 MHz (UHF)

International Table			United States Table		FCC Rule Part(s)
Region 1 Table	Region 2 Table	Region 3 Table	Federal Table	Non-Federal Table	
	698–806 MOBILE 5.313B 5.317A BROADCASTING Fixed		698–763	698–763 FIXED MOBILE BROADCASTING NG159	Wireless Communications (27) LPTV and TV Translator (74G)
			763–775	763–775 FIXED MOBILE NG158 NG159	Public Safety Land Mobile (90R)
			775–793	775–793 FIXED MOBILE BROADCASTING NG159	Wireless Communications (27) LPTV and TV Translator (74G)
790–862 FIXED MOBILE except aeronautical mobile 5.316B 5.317A BROADCASTING			793–805	793–805 FIXED MOBILE NG158 NG159	Public Safety Land Mobile (90R)

(Continued)

TABLE F.1 (Continued)

Worldwide Frequency Allocations

Table of Frequency Allocations

698–941 MHz (UHF)

International Table			United States Table		FCC Rule Part(s)
Region 1 Table	Region 2 Table	Region 3 Table	Federal Table	Non-Federal Table	
	5.293 5.309 5.311A		805–806	805–806 FIXED MOBILE BROADCASTING NG159	Wireless Communications (27) LPTV and TV Translator (74G)
	806–890 FIXED MOBILE 5.317A BROADCASTING		806–809	806–809 LAND MOBILE	Public Safety Land Mobile (90S)
			809–851	809–849 FIXED LAND MOBILE	Public Mobile (22) Private Land Mobile (90)
				849–851 AERONAUTICAL MOBILE	Public Mobile (22)
5.312 5.314 5.315 5.316 5.316A 5.319			851–854	851–854 LAND MOBILE	Public Safety Land Mobile (90S)

(Continued)

TABLE F.1 (Continued)
Worldwide Frequency Allocations
Table of Frequency Allocations

698–941 MHz (UHF)

International Table			United States Table		FCC Rule Part(s)
Region 1 Table	Region 2 Table	Region 3 Table	Federal Table	Non-Federal Table	
862–890 FIXED MOBILE except aeronautical mobile 5.317A BROADCASTING 5.322			854–890	854–894 FIXED LAND MOBILE	Public Mobile (22) Private Land Mobile (90)
5.319 5.323	5.317 5.318	5.317 5.318		US116 US268	
				894–896 AERONAUTICAL MOBILE	Public Mobile (22)
				US116 US268	
890–942 FIXED MOBILE except aeronautical mobile 5.317A BROADCASTING 5.322 Radiolocation	890–902 FIXED MOBILE except aeronautical mobile 5.317A Radiolocation	890–942 FIXED MOBILE 5.317A BROADCASTING Radiolocation	890–902	896–901 FIXED LAND MOBILE	Private Land Mobile (90)
				US116 US268	
				901–902 FIXED MOBILE	Personal Communications (24)
5.318 5.325	5.325		US116 US268 G2	US116 US268	

(Continued)

TABLE F.1 (*Continued*)

Worldwide Frequency Allocations

Table of Frequency Allocations   698–941 MHz (UHF)

International Table			United States Table			
Region 1 Table	**Region 2 Table**	**Region 3 Table**	**Federal Table**	**Non-Federal Table**	**FCC Rule Part(s)**	
	902–928 FIXED Amateur Mobile except aeronautical mobile 5.325A Radiolocation		902–928 RADIOLOCATION G59	902–928	ISM Equipment (18) Private Land Mobile (90) Amateur Radio (97)	
	5.150 5.325 5.326		5.150 US218 US267 US275 G11	5.150 US218 US267 US275		
	928–942 FIXED MOBILE except aeronautical mobile 5.317A Radiolocation		928–932	928–929 FIXED US116 US268 NG120	Public Mobile (22) Private Land Mobile (90) Fixed Microwave (101)	
				929–930 FIXED LAND MOBILE US116 US268	Private Land Mobile (90)	

(Continued)

TABLE F.1 (*Continued*)

Worldwide Frequency Allocations

Table of Frequency Allocations   698–941 MHz (UHF)

International Table			United States Table		FCC Rule Part(s)
Region 1 Table	Region 2 Table	Region 3 Table	Federal Table	Non-Federal Table	
				930–931 FIXED MOBILE	Personal Communications (24)
				US116 US268	
			US116 US268 G2	931–932 FIXED LAND MOBILE	Public Mobile (22)
				US116 US268	
			932–935 FIXED	932–935 FIXED	Public Mobile (22) Fixed Microwave (101)
			US268 G2	US268 NG120	
			935–941	935–940 FIXED LAND MOBILE	Private Land Mobile (90)
				US116 US268	
			US116 US268 G2	940–941 FIXED MOBILE	Personal Communications (24)
				US116 US268	
5.323	5.325	5.327			

(*Continued*)

TABLE F.1 (Continued)

Worldwide Frequency Allocations

Table of Frequency Allocations

941–1,525 MHz (UHF)

International Table			United States Table		FCC Rule Part(s)
Region 1 Table	Region 2 Table	Region 3 Table	Federal Table	Non-Federal Table	
942–960 FIXED MOBILE except aeronautical mobile 5.317A BROADCASTING 5.322	942–960 FIXED MOBILE 5.317A	942–960 FIXED MOBILE 5.317A BROADCASTING	941–944 FIXED US268 US301 G2	941–944 FIXED US268 US301 NG30 NG120	Public Mobile (22) Aural Broadcast Auxiliary (74E) Fixed Microwave (101)
			944–960	944–960 FIXED NG120	Public Mobile (22) Aural Broadcast Auxiliary (74E) Low Power Auxiliary (74H) Fixed Microwave (101)
5.323		5.320			
960–1164 AERONAUTICAL MOBILE (R) 5.327A AERONAUTICAL RADIONAVIGATION 5.328	AERONAUTICAL RADIONAVIGATION 5.328		960–1164 AERONAUTICAL RADIONAVIGATION 5.328 US224 US400		Aviation (87)
1164–1215 AERONAUTICAL RADIONAVIGATION 5.328 RADIONAVIGATION-SATELLITE (space-to-Earth) (space-to-space) 5.328B 5.328A			1164–1215 AERONAUTICAL RADIONAVIGATION 5.328 RADIONAVIGATION-SATELLITE (space-to-space) (space-to-Earth) 5.328A US224		

(Continued)

TABLE F.1 (*Continued*)

Worldwide Frequency Allocations

Table of Frequency Allocations

941–1,525 MHz (UHF)

International Table			United States Table		
Region 1 Table	Region 2 Table	Region 3 Table	Federal Table	Non-Federal Table	FCC Rule Part(s)
1215–1240 EARTH EXPLORATION-SATELLITE (active) RADIOLOCATION RADIONAVIGATION-SATELLITE (space-to-Earth) (space-to-space) 5.328B 5.329 5.329A SPACE RESEARCH (active)			1215–1240 EARTH EXPLORATION-SATELLITE (active) RADIOLOCATION G56 RADIO-NAVIGATION-SATELLITE (space-to-Earth) (space-to-space) G132 SPACE RESEARCH (active)	1215–1240 Earth exploration-satellite (active) Space research (active)	
5.330 5.331 5.332			5.332		

(*Continued*)

TABLE F.1 (Continued)

Worldwide Frequency Allocations

Table of Frequency Allocations			941–1,525 MHz (UHF)		
International Table			United States Table		
Region 1 Table	Region 2 Table	Region 3 Table	Federal Table	Non-Federal Table	FCC Rule Part(s)
1240–1300 EARTH EXPLORATION-SATELLITE (active) RADIOLOCATION RADIONAVIGATION-SATELLITE (space-to-Earth) (space-to-space) 5.328B 5.329 5.329A SPACE RESEARCH (active) Amateur			1240–1300 EARTH EXPLORATION-SATELLITE (active) RADIOLOCATION G56 SPACE RESEARCH (active) AERONAUTICAL RADIO-NAVIGATION	1240–1300 AERONAUTICAL RADIO-NAVIGATION Amateur Earth exploration-satellite (active) Space research (active)	Amateur Radio (97)
5.282 5.330 5.331 5.332 5.335 5.335A			5.332 5.335	5.282	
1300–1350 RADIOLOCATION AERONAUTICAL RADIONAVIGATION 5.337 RADIONAVIGATION-SATELLITE (Earth-to-space)			1300–1350 AERONAUTICAL RADIO-NAVIGATION 5.337 Radiolocation G2	1300–1350 AERONAUTICAL RADIO-NAVIGATION 5.337	Aviation (87)
5.149 5.337A			US342	US342	

(Continued)

TABLE F.1 (*Continued*)

Worldwide Frequency Allocations

Table of Frequency Allocations

941–1,525 MHz (UHF)

International Table			United States Table		FCC Rule Part(s)
Region 1 Table	Region 2 Table	Region 3 Table	Federal Table	Non-Federal Table	
1350–1400 FIXED MOBILE RADIOLOCATION	1350–1400 RADIOLOCATION 5.338A		1350–1390 FIXED MOBILE RADIOLOCATION G2	1350–1390	
			5.334 5.339 US342 US385 G27 G114	5.334 5.339 US342 US385	
			1390–1395	1390–1392 FIXED MOBILE except aeronautical mobile Fixed-satellite (Earth-to-space) US368	Wireless Communications (27)
				5.339 US37 US342 US385 US398	
				1392–1395 FIXED MOBILE except aeronautical mobile	
			5.339 US37 US342 US398	5.339 US37 US342 US385 US398	

(Continued)

TABLE F.1 (Continued)
Worldwide Frequency Allocations

Table of Frequency Allocations

941–1,525 MHz (UHF)

International Table			United States Table		FCC Rule Part(s)
Region 1 Table	Region 2 Table	Region 3 Table	Federal Table	Non-Federal Table	
			1395–1400 LAND MOBILE (medical telemetry and medical telecommand)	1395–1400 LAND MOBILE (medical telemetry and medical telecommand)	Personal Radio (95)
5.149 5.338 5.338A 5.339	5.149 5.334 5.339		5.339 US37 US342 US385 US398		
1400–1427 EARTH EXPLORATION-SATELLITE (passive) RADIO ASTRONOMY SPACE RESEARCH (passive) 5.340 5.341	EARTH EXPLORATION-SATELLITE (passive)		1400–1427 EARTH EXPLORATION-SATELLITE (passive) RADIO ASTRONOMY US74 SPACE RESEARCH (passive) 5.341 US246		
1427–1429 SPACE OPERATION (Earth-to-space) FIXED MOBILE except aeronautical mobile			1427–1429.5 LAND MOBILE (medical telemetry and medical telecommand) US350 5.341 US37 US398	1427–1429.5 LAND MOBILE (telemetry and telecommand) Fixed (telemetry) 5.341 US37 US350 US398	Private Land Mobile (90) Personal Radio (95)
5.338A 5.341					

(Continued)

TABLE F.1 (*Continued*)

Worldwide Frequency Allocations

Table of Frequency Allocations

941–1,525 MHz (UHF)

International Table			United States Table		FCC Rule Part(s)
Region 1 Table	Region 2 Table	Region 3 Table	Federal Table	Non-Federal Table	
1429–1452 FIXED MOBILE except aeronautical mobile	1429–1452 FIXED MOBILE 5.343		1429.5–1432	1429.5–1430 FIXED (telemetry and telecommand) LAND MOBILE (telemetry and telecommand)	
				5.341 US37 US350 US398	
				1430–1432 FIXED (telemetry and telecommand) LAND MOBILE (telemetry and telecommand) Fixed-satellite (space-to-Earth) US368	
			5.341 US37 US350 US398	5.341 US37 US350 US398	

(Continued)

TABLE F.1 (*Continued*)

Worldwide Frequency Allocations

Table of Frequency Allocations			941–1,525 MHz (UHF)		
International Table			United States Table		FCC Rule Part(s)
Region 1 Table	Region 2 Table	Region 3 Table	Federal Table	Non-Federal Table	
			1432–1435	1432–1435 FIXED MOBILE except aeronautical mobile	Wireless Communications (27)
			5.341 US83	5.341 US83	
5.338A 5.341 5.342	5.338A 5.341		1435–1525 MOBILE (aeronautical telemetry)		Aviation (87)
1452–1492 FIXED MOBILE except aeronautical mobile BROADCASTING 5.345 BROADCASTING-SATELLITE 5.208B 5.345	1452–1492 FIXED MOBILE 5.343 BROADCASTING 5.345 BROADCASTING-SATELLITE 5.208B 5.345				
5.341 5.342	5.341 5.344		5.341 US78		

(Continued)

TABLE F.1 (*Continued*)
Worldwide Frequency Allocations

Table of Frequency Allocations

1,525–1,670 MHz (UHF)

	International Table			United States Table		
Region 1 Table	Region 2 Table	Region 3 Table	Federal Table	Non-Federal Table	FCC Rule Part(s)
1492–1518 FIXED MOBILE except aeronautical mobile	1492–1518 FIXED MOBILE 5.343	1492–1518 FIXED MOBILE			
5.341 5.342	5.341 5.344	5.341			
1518–1525 FIXED MOBILE except aeronautical mobile MOBILE-SATELLITE (space-to-Earth) 5.348 5.348A 5.348B 5.351A 5.341 5.342	1518–1525 FIXED MOBILE 5.343 MOBILE-SATELLITE (space-to-Earth) 5.348 5.348A 5.348B 5.351A 5.341 5.344	1518–1525 FIXED MOBILE MOBILE-SATELLITE (space-to-Earth) 5.348 5.348A 5.348B 5.351A 5.341			

(*Continued*)

TABLE F.1 (*Continued*)

Worldwide Frequency Allocations

Table of Frequency Allocations 1,525–1,670 MHz (UHF)

International Table			United States Table		FCC Rule Part(s)
Region 1 Table	Region 2 Table	Region 3 Table	Federal Table	Non-Federal Table	
1525–1530 SPACE OPERATION (space-to-Earth) FIXED MOBILE-SATELLITE (space-to-Earth) 5.208B 5.351A Earth exploration-satellite Mobile except aeronautical mobile 5.349	1525–1530 SPACE OPERATION (space-to-Earth) MOBILE-SATELLITE (space-to-Earth) 5.208B 5.351A Earth exploration-satellite Fixed Mobile 5.343	1525–1530 SPACE OPERATION (space-to-Earth) FIXED MOBILE-SATELLITE (space-to-Earth) 5.208B 5.351A Earth exploration-satellite Mobile 5.349	1525–1535 MOBILE-SATELLITE (space-to-Earth) US380	MOBILE-SATELLITE (space-to-Earth) US315	Satellite Communications (25) Maritime (80)
5.341 5.342 5.350 5.351 5.352A 5.354	5.341 5.351 5.354	5.341 5.351 5.352A 5.354			

(*Continued*)

TABLE F.1 (*Continued*)

Worldwide Frequency Allocations

Table of Frequency Allocations   1,525–1,670 MHz (UHF)

International Table			United States Table		FCC Rule Part(s)
Region 1 Table	Region 2 Table	Region 3 Table	Federal Table	Non-Federal Table	
1530–1535 SPACE OPERATION (space-to-Earth) MOBILE-SATELLITE (space-to-Earth) 5.208B 5.351A 5.353A Earth exploration-satellite Fixed Mobile except aeronautical mobile	1530–1535 SPACE OPERATION (space-to-Earth) MOBILE-SATELLITE (space-to-Earth) 5.208B 5.351A 5.353A Earth exploration-satellite Fixed Mobile 5.343				
5.341 5.342 5.351 5.354	5.341 5.351 5.354		5.341 5.351		
1535–1559 MOBILE-SATELLITE (space-to-Earth) 5.208B 5.351A			1535–1559 MOBILE-SATELLITE (space-to-Earth) US308 US309 US315 US380		Satellite Communications (25) Maritime (80) Aviation (87)
5.341 5.351 5.353A 5.354 5.355 5.356 5.357 5.357A 5.359 5.362A			5.341 5.351 5.356		

(Continued)

TABLE F.1 (*Continued*)

Worldwide Frequency Allocations

Table of Frequency Allocations   1,525–1,670 MHz (UHF)

International Table			United States Table		FCC Rule Part(s)
Region 1 Table	Region 2 Table	Region 3 Table	Federal Table	Non-Federal Table	
1559–1610 AERONAUTICAL RADIONAVIGATION RADIONAVIGATION-SATELLITE (space-to-Earth) (space-to-space) 5.208B 5.328B 5.329A	AERONAUTICAL RADIONAVIGATION RADIONAVIGATION-SATELLITE (space-to-Earth) (space-to-space)		1559–1610 AERONAUTICAL RADIONAVIGATION RADIONAVIGATION-SATELLITE (space-to-Earth) (space-to-space) 5.341 US208 US260 US343	AERONAUTICAL RADIONAVIGATION RADIONAVIGATION-SATELLITE	Aviation (87)
5.341 5.362B 5.362C					
1610–1610.6 MOBILE-SATELLITE (Earth-to-space) 5.351A AERONAUTICAL RADIONAVIGATION	1610–1610.6 MOBILE-SATELLITE (Earth-to-space) 5.351A AERONAUTICAL RADIO-NAVIGATION RADIO-DETERMINATION-SATELLITE (Earth-to-space)	1610–1610.6 MOBILE-SATELLITE (Earth-to-space) 5.351A AERONAUTICAL RADIO-NAVIGATION Radiodetermination-satellite (Earth-to-space)	1610–1610.6 MOBILE-SATELLITE (Earth-to-space) US380 AERONAUTICAL RADIONAVIGATION US260 RADIODETERMINATION-SATELLITE (Earth-to-space)	1610–1610.6 MOBILE-SATELLITE (Earth-to-space) US319 AERONAUTICAL RADIONAVIGATION RADIODETERMINATION-SATELLITE	Satellite Communications (25) Aviation (87)
5.341 5.355 5.359 5.364 5.366 5.367 5.368 5.369 5.371 5.372	5.341 5.364 5.366 5.367 5.368 5.370 5.372	5.341 5.355 5.359 5.364 5.366 5.367 5.368 5.369 5.372		5.341 5.364 5.366 5.367 5.368 5.372 US208	

(*Continued*)

TABLE F.1 (*Continued*)

Worldwide Frequency Allocations

Table of Frequency Allocations

1,525–1,670 MHz (UHF)

International Table			United States Table		FCC Rule Part(s)
Region 1 Table	Region 2 Table	Region 3 Table	Federal Table	Non-Federal Table	
1610.6–1613.8 MOBILE-SATELLITE (Earth-to-space) 5.351A RADIO ASTRONOMY AERONAUTICAL RADIONAVIGATION	1610.6–1613.8 MOBILE-SATELLITE (Earth-to-space) 5.351A RADIO ASTRONOMY AERONAUTICAL RADIO-NAVIGATION RADIO-DETERMINATION-SATELLITE (Earth-to-space) 5.149 5.341 5.364 5.366 5.367 5.368 5.370 5.372	1610.6–1613.8 MOBILE-SATELLITE (Earth-to-space) 5.351A RADIO ASTRONOMY AERONAUTICAL RADIO-NAVIGATION Radiodetermination-satellite (Earth-to-space) 5.149 5.341 5.355 5.359 5.364 5.366	1610.6–1613.8 MOBILE-SATELLITE (Earth-to-space) US380 RADIO ASTRONOMY AERONAUTICAL RADIONAVIGATION US260 RADIODETERMINATION-SATELLITE (Earth-to-space)	1610.6–1613.8 MOBILE-SATELLITE (Earth-to-space) US319	
5.149 5.341 5.355 5.359 5.364 5.366 5.367 5.368 5.369 5.371 5.372		5.367 5.368 5.369 5.372	5.341 5.364 5.366 5.367 5.368 US342	5.366 5.367 5.368 5.372 US208	

(*Continued*)

TABLE F.1 (*Continued*)

Worldwide Frequency Allocations

Table of Frequency Allocations

1,525–1,670 MHz (UHF)

International Table			United States Table		FCC Rule Part(s)
Region 1 Table	Region 2 Table	Region 3 Table	Federal Table	Non-Federal Table	
1613.8–1626.5 MOBILE-SATELLITE (Earth-to-space) 5.351A AERONAUTICAL RADIONAVIGATION Mobile-satellite (space-to-Earth) 5.208B	1613.8–1626.5 MOBILE-SATELLITE (Earth-to-space) 5.351A AERONAUTICAL RADIO-NAVIGATION RADIO-DETERMINATION-SATELLITE (Earth-to-space) Mobile-satellite (space-to-Earth) 5.208B	1613.8–1626.5 MOBILE-SATELLITE (Earth-to-space) 5.351A AERONAUTICAL RADIO-NAVIGATION Mobile-satellite (space-to-Earth) 5.208B Radiodetermination-satellite (Earth-to-space)	1613.8–1626.5 MOBILE-SATELLITE (Earth-to-space) US380 AERONAUTICAL RADIONAVIGATION US260 RADIODETERMINATION-SATELLITE (Earth-to-space) Mobile-satellite (space-to-Earth)	1613.8–1626.5 MOBILE-SATELLITE (Earth-to-space) US319 AERONAUTICAL RADIONAVIGATION RADIODETERMINATION-SATELLITE (space-to-Earth)	
5.341 5.355 5.359 5.364 5.365 5.366 5.367 5.368 5.369 5.371 5.372	5.341 5.364 5.365 5.366 5.367 5.368 5.370 5.372	5.341 5.355 5.359 5.364 5.365 5.366 5.367 5.368 5.369 5.372	5.341 5.364 5.365 5.366 5.367 5.368 5.372 US208		
1626.5–1660 MOBILE-SATELLITE (Earth-to-space) 5.351A			1626.5–1660 MOBILE-SATELLITE (Earth-to-space) US309 US315 US380	1626.5–1660 MOBILE-SATELLITE (Earth-to-space) US308	Satellite Communications (25) Maritime (80) Aviation (87)
5.341 5.351 5.353A 5.354 5.355 5.357A 5.359 5.362A 5.374 5.375 5.376			5.341 5.351 5.375		

(Continued)

TABLE F.1 (*Continued*)

Worldwide Frequency Allocations

Table of Frequency Allocations

1,525–1,670 MHz (UHF)

International Table			United States Table		FCC Rule Part(s)
Region 1 Table	Region 2 Table	Region 3 Table	Federal Table	Non-Federal Table	
1660–1660.5 MOBILE-SATELLITE (Earth-to-space) 5.351A RADIO ASTRONOMY			1660–1660.5 MOBILE-SATELLITE (Earth-to-space) US308 US309 US380 RADIO ASTRONOMY	1660–1660.5 MOBILE-SATELLITE (Earth-to-space) US309 RADIO ASTRONOMY	Satellite Communications (25) Aviation (87)
5.149 5.341 5.351 5.354 5.362A 5.376A			5.341 5.351 US342		
1660.5–1668 RADIO ASTRONOMY SPACE RESEARCH (passive) Fixed Mobile except aeronautical mobile			1660.5–1668.4 RADIO ASTRONOMY US74 SPACE RESEARCH (passive)		
5.149 5.341 5.379 5.379A					
1668–1668.4 MOBILE-SATELLITE (Earth-to-space) 5.351A 5.379B 5.379C RADIO ASTRONOMY SPACE RESEARCH (passive) Fixed Mobile except aeronautical mobile					
5.149 5.341 5.379 5.379A			5.341 US246		

(Continued)

TABLE F.1 (*Continued*)

Worldwide Frequency Allocations

Table of Frequency Allocations

1,525–1,670 MHz (UHF)

International Table			United States Table		FCC Rule Part(s)
Region 1 Table	Region 2 Table	Region 3 Table	Federal Table	Non-Federal Table	
1668.4–1670 METEOROLOGICAL AIDS FIXED MOBILE except aeronautical mobile MOBILE-SATELLITE (Earth-to-space) 5.351A 5.379B 5.379C RADIO ASTRONOMY			1668.4–1670 METEOROLOGICAL AIDS (radiosonde) RADIO ASTRONOMY US74		
5.149 5.341 5.379D 5.379E			5.341 US99 US342		

Table of Frequency Allocations

1,670–2,200 MHz (UHF)

International Table			United States Table		FCC Rule Part(s)
Region 1 Table	Region 2 Table	Region 3 Table	Federal Table	Non-Federal Table	
1670–1675 METEOROLOGICAL AIDS FIXED METEOROLOGICAL-SATELLITE (space-to-Earth) MOBILE MOBILE-SATELLITE (Earth-to-space) 5.351A 5.379B			1670–1675	1670–1675 FIXED MOBILE except aeronautical mobile	Wireless Communications (27)
5.341 5.379D 5.379E 5.380A			5.341 US211 US362	5.341 US211 US362	
1675–1690 METEOROLOGICAL AIDS FIXED METEOROLOGICAL-SATELLITE (space-to-Earth) MOBILE except aeronautical mobile 5.341			1675–1695 METEOROLOGICAL AIDS (radiosonde) METEOROLOGICAL-SATELLITE (space-to-Earth) US88		
			5.341 US211 US289		

(*Continued*)

TABLE F.1 (*Continued*)
Worldwide Frequency Allocations
Table of Frequency Allocations 1,670–2,200 MHz (UHF)

International Table			United States Table		FCC Rule Part(s)
Region 1 Table	Region 2 Table	Region 3 Table	Federal Table	Non-Federal Table	
1690–1700 METEOROLOGICAL AIDS METEOROLOGICAL-SATELLITE (space-to-Earth) Fixed Mobile except aeronautical mobile	1690–1700 METEOROLOGICAL AIDS METEOROLOGICAL-SATELLITE (space-to-Earth)		1695–1710 METEORO-LOGICAL-SATELLITE (space-to-Earth) US88	1695–1710 FIXED MOBILE except aeronautical mobile	Wireless Communications (27)
5.289 5.341 5.382	5.289 5.341 5.381				
1700–1710 FIXED METEOROLOGICAL-SATELLITE (space-to-Earth) MOBILE except aeronautical mobile		1700–1710 FIXED METEOROLOGICAL-SATELLITE (space-to-Earth) MOBILE except aeronautical mobile			
5.289 5.341		5.289 5.341 5.384	5.341	5.341 US88	

(*Continued*)

TABLE F.1 (*Continued*)
Worldwide Frequency Allocations

Table of Frequency Allocations			1,670–2,200 MHz (UHF)		
International Table			United States Table		FCC Rule Part(s)
Region 1 Table	Region 2 Table	Region 3 Table	Federal Table	Non-Federal Table	
1710–1930 FIXED MOBILE 5.384A 5.388A 5.388B			1710–1761	1710–1780 FIXED MOBILE	
			5.341 US91 US378 US385		
			1761–1780 SPACE OPERATION (Earth-to-space) G42 US91	5.341 US91 US378 US385	
			1780–1850 FIXED MOBILE SPACE OPERATION (Earth-to-space) G42	1780–1850	
5.149 5.341 5.385 5.386 5.387 5.388			1850–2025	1850–2000 FIXED MOBILE	RF Devices (15) Personal Communications (24) Wireless Communications (27) Fixed Microwave (101)
1930–1970 FIXED MOBILE 5.388A 5.388B	1930–1970 FIXED MOBILE 5.388A 5.388B Mobile-satellite (Earth-to-space) 5.388	1930–1970 FIXED MOBILE 5.388A 5.388B			
5.388	5.388	5.388			

(*Continued*)

TABLE F.1 (Continued)

Worldwide Frequency Allocations

Table of Frequency Allocations

1,670–2,200 MHz (UHF)

International Table			United States Table		
Region 1 Table	Region 2 Table	Region 3 Table	Federal Table	Non-Federal Table	FCC Rule Part(s)
1970–1980 FIXED MOBILE 5.388A 5.388B					
5.388					
1980–2010 FIXED MOBILE MOBILE-SATELLITE (Earth-to-space) 5.351A 5.388 5.389A 5.389B 5.389F				2000–2020 FIXED MOBILE MOBILE-SATELLITE (Earth-to-space)	Satellite Communications (25) Wireless Communications (27)
2010–2025 FIXED MOBILE 5.388A 5.388B	2010–2025 FIXED MOBILE MOBILE-SATELLITE (Earth-to-space)	2010–2025 FIXED MOBILE 5.388A 5.388B		2020–2025 FIXED MOBILE	
5.388	5.388 5.389C 5.389E	5.388			

(Continued)

TABLE F.1 (Continued)

Worldwide Frequency Allocations

Table of Frequency Allocations

1,670–2,200 MHz (UHF)

International Table			United States Table		FCC Rule Part(s)
Region 1 Table	Region 2 Table	Region 3 Table	Federal Table	Non-Federal Table	
2025–2110 SPACE OPERATION (Earth-to-space) (space-to-space) EARTH EXPLORATION-SATELLITE (Earth-to-space) (space-to-space) FIXED MOBILE 5.391 SPACE RESEARCH (Earth-to-space) (space-to-space) 5.392	SPACE OPERATION (Earth-to-space) (space-to-space) EARTH EXPLORATION-SATELLITE (Earth-to-space) (space-to-space) SPACE RESEARCH (Earth-to-space) (space-to-space)		2025–2110 SPACE OPERATION (Earth-to-space) (space-to-space) EARTH EXPLORATION-SATELLITE (Earth-to-space) (space-to-space) FIXED MOBILE 5.391 SPACE RESEARCH (Earth-to-space) (space-to-space) 5.392 US90 US92 US222 US346 US347	2025–2110 FIXED NG118 MOBILE 5.391 5.392 US90 US92 US222 US346 US347	TV Auxiliary Broadcasting (74F) Cable TV Relay (78) Local TV Transmission (101J)
2110–2120 FIXED MOBILE 5.388A 5.388B SPACE RESEARCH (deep space) (Earth-to-space) 5.388			2110–2120 US252	2110–2120 FIXED MOBILE US252	Public Mobile (22) Wireless Communications (27) Fixed Microwave (101)

(Continued)

TABLE F.1 (Continued)

Worldwide Frequency Allocations

Table of Frequency Allocations　　1,670–2,200 MHz (UHF)

International Table			United States Table		FCC Rule Part(s)
Region 1 Table	Region 2 Table	Region 3 Table	Federal Table	Non-Federal Table	
2120–2170 FIXED MOBILE 5.388A 5.388B	2120–2160 FIXED MOBILE 5.388A 5.388B Mobile-satellite (space-to-Earth) 5.388B 5.388 2160–2170 FIXED MOBILE MOBILE-SATELLITE (space-to-Earth) 5.388 5.389C 5.389E	2120–2170 FIXED MOBILE 5.388A 5.388B 5.388	2120–2200	2120–2180 FIXED MOBILE	
5.388				NG41	
2170–2200 FIXED MOBILE MOBILE-SATELLITE (space-to-Earth) 5.351A 5.388 5.389A 5.389F	MOBILE-SATELLITE (space-to-Earth) 5.351A			2180–2200 FIXED MOBILE MOBILE-SATELLITE (space-to-Earth)	Satellite Communications (25) Wireless Communications (27)

(Continued)

TABLE F.1 (Continued)

Worldwide Frequency Allocations

Table of Frequency Allocations

2,200–2,655 MHz (UHF)

International Table			United States Table		FCC Rule Part(s)
Region 1 Table	Region 2 Table	Region 3 Table	Federal Table	Non-Federal Table	
2200–2290			2200–2290	2200–2290	
SPACE OPERATION (space-to-Earth) (space-to-space)			SPACE OPERATION (space-to-Earth) (space-to-space)		
EARTH EXPLORATION-SATELLITE (space-to-Earth) (space-to-space)			EARTH EXPLORATION -SATELLITE (space-to-Earth) (space-to-space)		
FIXED			FIXED (line-of-sight only)		
MOBILE 5.391			MOBILE (line-of-sight only including aeronautical telemetry, but excluding flight testing of manned aircraft) 5.391	US303	
SPACE RESEARCH (space-to-Earth) (space-to-space)			SPACE RESEARCH (space-to-Earth) (space-to-space)		
5.392			5.392 US303		

(*Continued*)

TABLE F.1 (*Continued*)

Worldwide Frequency Allocations

Table of Frequency Allocations 2,200–2,655 MHz (UHF)

International Table			United States Table		FCC Rule Part(s)
Region 1 Table	Region 2 Table	Region 3 Table	Federal Table	Non-Federal Table	
2290–2300 FIXED MOBILE except aeronautical mobile SPACE RESEARCH (deep space) (space-to-Earth)	MOBILE except aeronautical mobile SPACE RESEARCH (deep space) (space-to-Earth)		2290–2300 FIXED MOBILE except aeronautical mobile SPACE RESEARCH (deep space) (space-to-Earth)	2290–2300 SPACE RESEARCH (deep space) (space-to-Earth)	
2300–2450 FIXED MOBILE 5.384A Amateur Radiolocation	2300–2450 FIXED MOBILE 5.384A RADIOLOCATION Amateur		2300–2305 G122	2300–2305 Amateur	Amateur Radio (97)
			2305–2310 US97 G122	2305–2310 FIXED MOBILE except aeronautical mobile RADIOLOCATION Amateur US97	Wireless Communications (27) Amateur Radio (97)

(Continued)

TABLE F.1 (*Continued*)
Worldwide Frequency Allocations

Table of Frequency Allocations   2,200–2,655 MHz (UHF)

International Table			United States Table		
Region 1 Table	Region 2 Table	Region 3 Table	Federal Table	Non-Federal Table	FCC Rule Part(s)
			2310–2320 Fixed Mobile US339 Radiolocation G2	2310–2320 FIXED MOBILE US339 BROADCASTING- SATELLITE RADIOLOCATION	Wireless Communications (27) Aviation (87)
			US97 US327	5.396 US97 US327	
			2320–2345 Fixed Radiolocation G2	2320–2345 BROADCASTING- SATELLITE	Satellite Communications (25)
			US327	5.396 US327	
			2345–2360 Fixed Mobile US339 Radiolocation G2	2345–2360 FIXED MOBILE US339 BROADCASTING- SATELLITE RADIOLOCATION	Wireless Communications (27) Aviation (87)
			US327	5.396 US327	

(*Continued*)

TABLE F.1 (Continued)

Worldwide Frequency Allocations

Table of Frequency Allocations 2,200–2,655 MHz (UHF)

International Table			United States Table		FCC Rule Part(s)
Region 1 Table	Region 2 Table	Region 3 Table	Federal Table	Non-Federal Table	
			2360–2390 MOBILE US276 RADIOLOCATION G2 G120 Fixed	2360–2390 MOBILE US276	Aviation (87) Personal Radio (95)
			US101	US101	
			2390–2395 MOBILE US276	2390–2395 AMATEUR MOBILE US276	Aviation (87) Personal Radio (95)
			US101	US101	Amateur Radio (97)
			2395–2400	2395–2400 AMATEUR	
			US101 G122	US101	Personal Radio (95) Amateur Radio (97)
			2400–2417	2400–2417 AMATEUR	
			5.150 G122	5.150 5.282	ISM Equipment (18) Amateur Radio (97)
			2417–2450 Radiolocation G2	2417–2450 Amateur	
5.150 5.282 5.395	5.150 5.282 5.393 5.394 5.396		5.150	5.150 5.282	

(Continued)

TABLE F.1 (*Continued*)

Worldwide Frequency Allocations

Table of Frequency Allocations   2,200–2,655 MHz (UHF)

International Table			United States Table		FCC Rule Part(s)
Region 1 Table	Region 2 Table	Region 3 Table	Federal Table	Non-Federal Table	
2450–2483.5 FIXED MOBILE Radiolocation 5.150 5.397	2450–2483.5 FIXED MOBILE RADIOLOCATION 5.150		2450–2483.5 5.150 US41	2450–2483.5 FIXED MOBILE Radiolocation 5.150 US41	ISM Equipment (18) TV Auxiliary Broadcasting (74F) Private Land Mobile (90) Fixed Microwave (101)
2483.5–2500 FIXED MOBILE MOBILE-SATELLITE (space-to-Earth) 5.351A Radiolocation	2483.5–2500 FIXED MOBILE MOBILE-SATELLITE (space-to-Earth) 5.351A RADIO-DETERMINATION-SATELLITE (space-to-Earth) 5.398 RADIOLOCATION	2483.5–2500 FIXED MOBILE MOBILE-SATELLITE (space-to-Earth) 5.351A RADIOLOCATION Radiodetermination-satellite (space-to-Earth) 5.398	2483.5–2500 MOBILE-SATELLITE (space-to-Earth) US319 US380 US391 RADIO-DETERMINATION-SATELLITE (space-to-Earth) 5.398	2483.5–2495 MOBILE-SATELLITE (space-to-Earth) US380 RADIO-DETERMINATION-SATELLITE (space-to-Earth) 5.398 5.150 5.402 US41 US319 NG147	ISM Equipment (18) Satellite Communications (25)

(Continued)

TABLE F.1 (*Continued*)

Worldwide Frequency Allocations

Table of Frequency Allocations   2,200–2,655 MHz (UHF)

International Table			United States Table		FCC Rule Part(s)
Region 1 Table	Region 2 Table	Region 3 Table	Federal Table	Non-Federal Table	
				2495–2500 FIXED MOBILE except aeronautical mobile MOBILE-SATELLITE (space-to-Earth) US380 RADIO-DETERMINATION-SATELLITE (space-to-Earth) 5.398	ISM Equipment (18) Satellite Communications (25) Wireless Communications (27)
5.150 5.371 5.397 5.398 5.399 5.400 5.402	5.150 5.402	5.150 5.400 5.402	5.150 5.402 US41	5.150 5.402 US41 US319 US391 NG147	

(Continued)

TABLE F.1 (*Continued*)

Worldwide Frequency Allocations

Table of Frequency Allocations

2,200–2,655 MHz (UHF)

	International Table			United States Table		
Region 1 Table	Region 2 Table	Region 3 Table		Federal Table	Non-Federal Table	FCC Rule Part(s)
2500–2520 FIXED 5.410 MOBILE except aeronautical mobile 5.384A	2500–2520 FIXED 5.410 FIXED-SATELLITE (space-to-Earth) 5.415 MOBILE except aeronautical mobile 5.384A	2500–2520 FIXED 5.410 FIXED-SATELLITE (space-to-Earth) 5.415 MOBILE except aeronautical mobile 5.384A MOBILE-SATELLITE (space-to-Earth) 5.351A 5.407 5.414 5.414A		2500–2655	2500–2655 FIXED US205 MOBILE except aeronautical mobile	Wireless Communications (27)
5.405 5.412	5.404	5.404 5.415A				

(*Continued*)

TABLE F.1 (Continued)

Worldwide Frequency Allocations

Table of Frequency Allocations 2,200–2,655 MHz (UHF)

International Table			United States Table		FCC Rule Part(s)
Region 1 Table	Region 2 Table	Region 3 Table	Federal Table	Non-Federal Table	
2520–2655 FIXED 5.410 MOBILE except aeronautical mobile 5.384A BROADCASTING-SATELLITE 5.413 5.416	2520–2655 FIXED 5.410 FIXED-SATELLITE (space-to-Earth) 5.415 MOBILE except aeronautical mobile 5.384A BROADCASTING-SATELLITE 5.413 5.416	2520–2535 FIXED 5.410 FIXED-SATELLITE (space-to-Earth) 5.415 MOBILE except aeronautical mobile 5.384A BROADCASTING-SATELLITE 5.413 5.416 5.403 5.414A 5.415A	5.339 US205	5.339	
		2535–2655 FIXED 5.410 MOBILE except aeronautical mobile 5.384A BROADCASTING-SATELLITE 5.413 5.416 5.339 5.417A 5.417B 5.417C 5.417D 5.418 5.418A 5.418B 5.418C			
5.339 5.405 5.412 5.417C 5.417D 5.418B 5.418C	5.339 5.417C 5.417D 5.418B 5.418C				

(Continued)

TABLE F.1 (Continued)

Worldwide Frequency Allocations

Table of Frequency Allocations   **2,655–4,990 MHz (UHF/SHF)**

International Table			United States Table		FCC Rule Part(s)
Region 1 Table	Region 2 Table	Region 3 Table	Federal Table	Non-Federal Table	
2655–2670 FIXED 5.410 MOBILE except aeronautical mobile 5.384A BROADCASTING-SATELLITE 5.208B 5.413 5.416 Earth exploration-satellite (passive) Radio astronomy Space research (passive)	2655–2670 FIXED 5.410 FIXED-SATELLITE (Earth-to-space) 5.415 (space-to-Earth) 5.415 MOBILE except aeronautical mobile 5.384A BROADCASTING-SATELLITE 5.413 5.416 Earth exploration-satellite (passive) Radio astronomy Space research (passive)	2655–2670 FIXED 5.410 FIXED-SATELLITE (Earth-to-space) 5.415 MOBILE except aeronautical mobile 5.384A BROADCASTING-SATELLITE 5.413 5.416 Earth exploration-satellite (passive) Radio astronomy Space research (passive)	2655–2690 Earth exploration-satellite (passive) Radio astronomy US385 Space research (passive)	2655–2690 FIXED US205 MOBILE except aeronautical mobile Earth exploration-satellite (passive) Radio astronomy Space research (passive)	Wireless Communications (27)
5.149 5.412	5.149 5.208B	5.149 5.208B 5.420			

(Continued)

TABLE F.1 (Continued)

Worldwide Frequency Allocations

Table of Frequency Allocations — 2,655–4,990 MHz (UHF/SHF)

International Table			United States Table		FCC Rule Part(s)
Region 1 Table	Region 2 Table	Region 3 Table	Federal Table	Non-Federal Table	
2670–2690 FIXED 5.410 MOBILE except aeronautical mobile 5.384A Earth exploration-satellite (passive) Radio astronomy Space research (passive)	2670–2690 FIXED 5.410 FIXED-SATELLITE (Earth-to-space) (space-to-Earth) 5.208B 5.415 MOBILE except aeronautical mobile 5.384A Earth exploration-satellite (passive) Radio astronomy Space research (passive)	2670–2690 FIXED 5.410 FIXED-SATELLITE (Earth-to-space) 5.415 MOBILE except aeronautical mobile 5.384A MOBILE-SATELLITE (Earth-to-space) 5.351A 5.419 Earth exploration-satellite (passive) Radio astronomy Space research (passive)	US205	US385	
5.149 5.412	5.149	5.149			
2690–2700 EARTH EXPLORATION-SATELLITE (passive) RADIO ASTRONOMY SPACE RESEARCH (passive)	EARTH EXPLORATION-SATELLITE (passive) RADIO ASTRONOMY SPACE RESEARCH (passive)		2690–2700 EARTH EXPLORATION-SATELLITE (passive) RADIO ASTRONOMY US74 SPACE RESEARCH (passive)	EARTH EXPLORATION-SATELLITE (passive)	
5.340 5.422			US246		

(Continued)

TABLE F.1 (*Continued*)

Worldwide Frequency Allocations

Table of Frequency Allocations

2,655–4,990 MHz (UHF/SHF)

International Table			United States Table		FCC Rule Part(s)
Region 1 Table	Region 2 Table	Region 3 Table	Federal Table	Non-Federal Table	
2700–2900 AERONAUTICAL RADIONAVIGATION 5.337 Radiolocation			2700–2900 METEOROLOGICAL AIDS AERONAUTICAL RADIONAVIGATION 5.337 US18 Radiolocation G2	2700–2900	Aviation (87)
5.423 5.424			5.423 G15	5.423 US18	
2900–3100 RADIOLOCATION 5.424A RADIONAVIGATION 5.426			2900–3100 RADIOLOCATION 5.424A G56 MARITIME RADIO-NAVIGATION	2900–3100 MARITIME RADIO-NAVIGATION Radiolocation US44	Maritime (80) Private Land Mobile (90)
5.425 5.427			5.427 US44 US316	5.427 US316	
3100–3300 RADIOLOCATION Earth exploration-satellite (active) Space research (active)			3100–3300 RADIOLOCATION G59 Earth exploration-satellite (active) Space research (active)	3100–3300 Earth exploration-satellite (active) Space research (active) Radiolocation	Private Land Mobile (90)
5.149 5.428			US342	US342	

(Continued)

TABLE F.1 (Continued)
Worldwide Frequency Allocations

Table of Frequency Allocations 2,655–4,990 MHz (UHF/SHF)

	International Table			United States Table		FCC Rule Part(s)
	Region 1 Table	Region 2 Table	Region 3 Table	Federal Table	Non-Federal Table	
	3300–3400 RADIOLOCATION	3300–3400 RADIOLOCATION Amateur Fixed Mobile	3300–3400 RADIOLOCATION Amateur	3300–3500 RADIOLOCATION US108 G2	3300–3500 Amateur Radiolocation US108	Private Land Mobile (90) Amateur Radio (97)
	5.149 5.429 5.430	5.149	5.149 5.429			
	3400–3600 FIXED FIXED-SATELLITE (space-to-Earth) Mobile 5.430A Radiolocation	3400–3500 FIXED FIXED-SATELLITE (space-to-Earth) Amateur Mobile 5.431A Radiolocation 5.433	3400–3500 FIXED FIXED-SATELLITE (space-to-Earth) Amateur Mobile 5.432B Radiolocation 5.433			
		5.282	5.282 5.432 5.432A	US342	5.282 US342	
		3500–3700 FIXED FIXED-SATELLITE (space-to-Earth) MOBILE except aeronautical mobile Radiolocation 5.433	3500–3600 FIXED FIXED-SATELLITE (space-to-Earth) MOBILE except aeronautical mobile 5.433A Radiolocation 5.433	3500–3650 RADIOLOCATION G59 AERONAUTICAL RADIO-NAVIGATION (ground-based) G110	3500–3600 Radiolocation	Private Land Mobile (90)
	5.431					

(Continued)

TABLE F.1 (*Continued*)

Worldwide Frequency Allocations

Table of Frequency Allocations					2,655–4,990 MHz (UHF/SHF)
International Table			**United States Table**		**FCC Rule Part(s)**
Region 1 Table	Region 2 Table	Region 3 Table	Federal Table	Non-Federal Table	
3600–4200 FIXED FIXED-SATELLITE (space-to-Earth) Mobile		3600–3700 FIXED FIXED-SATELLITE (space-to-Earth) MOBILE except aeronautical mobile Radiolocation 5.433	US245	3600–3650 FIXED-SATELLITE (space-to-Earth) US245 Radiolocation	Satellite Communications (25) Private Land Mobile (90)
		5.435	3650–3700	3650–3700 FIXED FIXED-SATELLITE (space-to-Earth) NG169 NG185 MOBILE except aeronautical mobile	
			US109 US349	US109 US349	
	3700–4200 FIXED FIXED-SATELLITE (space-to-Earth) MOBILE except aeronautical mobile		3700–4200	3700–4200 FIXED FIXED-SATELLITE (space-to-Earth) NG180	Satellite Communications (25) Fixed Microwave (101)
4200–4400 AERONAUTICAL RADIONAVIGATION 5.439 5.440	AERONAUTICAL RADIONAVIGATION 5.438		4200–4400 AERONAUTICAL RADIONAVIGATION 5.440 US261	AERONAUTICAL RADIONAVIGATION	Aviation (87)

(Continued)

TABLE F.1 (Continued)

Worldwide Frequency Allocations

Table of Frequency Allocations — 2,655–4,990 MHz (UHF/SHF)

International Table			United States Table		FCC Rule Part(s)
Region 1 Table	Region 2 Table	Region 3 Table	Federal Table	Non-Federal Table	
4400–4500 FIXED MOBILE 5.440A			4400–4500 FIXED MOBILE	4400–4500	
4500–4800 FIXED FIXED-SATELLITE (space-to-Earth) 5.441 MOBILE 5.440A			4500–4800 FIXED MOBILE US245	4500–4800 FIXED-SATELLITE (space-to-Earth) 5.441 US245	
4800–4990 FIXED MOBILE 5.440A 5.442 Radio astronomy			4800–4940 FIXED MOBILE US203 US342	4800–4940 US203 US342	
			4940–4990	4940–4990 FIXED MOBILE except aeronautical mobile	Public Safety Land Mobile (90Y)
5.149 5.339 5.443			5.339 US342 US385 G122	5.339 US342 US385	

(Continued)

TABLE F.1 (*Continued*)

Worldwide Frequency Allocations

Table of Frequency Allocations — 4,990–5,925 MHz (SHF)

International Table			United States Table		FCC Rule Part(s)
Region 1 Table	Region 2 Table	Region 3 Table	Federal Table	Non-Federal Table	
4990–5000 FIXED MOBILE except aeronautical mobile RADIO ASTRONOMY Space research (passive)			4990–5000 RADIO ASTRONOMY US74 Space research (passive)		
5.149			US246		
5000–5010 AERONAUTICAL RADIONAVIGATION RADIONAVIGATION-SATELLITE (Earth-to-space)			5000–5010 AERONAUTICAL RADIONAVIGATION US260 RADIONAVIGATION-SATELLITE (Earth-to-space)	AERONAUTICAL RADIONAVIGATION	Aviation (87)
5.367			5.367 US211		
5010–5030 AERONAUTICAL RADIONAVIGATION RADIONAVIGATION-SATELLITE (space-to-Earth) (space-to-space) 5.328B 5.443B			5010–5030 AERONAUTICAL RADIONAVIGATION US260 RADIONAVIGATION-SATELLITE (space-to-Earth) (space-to-space) 5.443B	AERONAUTICAL RADIONAVIGATION RADIONAVIGATION-SATELLITE (space-to-Earth) 5.443B	
5.367			5.367 US211		

(*Continued*)

TABLE F.1 (*Continued*)

Worldwide Frequency Allocations

Table of Frequency Allocations			4,990–5,925 MHz (SHF)		
International Table			United States Table		FCC Rule Part(s)
Region 1 Table	Region 2 Table	Region 3 Table	Federal Table	Non-Federal Table	
5030–5091 AERONAUTICAL RADIONAVIGATION			5030–5091 AERONAUTICAL RADIONAVIGATION US260	AERONAUTICAL RADIONAVIGATION	
5.367 5.444			5.367 US211 US444		
5091–5150 AERONAUTICAL RADIONAVIGATION AERONAUTICAL MOBILE 5.444B			5091–5150 AERONAUTICAL RADIONAVIGATION US260	AERONAUTICAL RADIONAVIGATION	Satellite Communications (25) Aviation (87)
5.367 5.444 5.444A			5.367 US211 US344 US444 US444A		
5150–5250 AERONAUTICAL RADIONAVIGATION FIXED-SATELLITE (Earth-to-space) 5.447A MOBILE except aeronautical mobile 5.446A 5.446B			5150–5250 AERONAUTICAL RADIO-NAVIGATION US260	5150–5250 AERONAUTICAL RADIO-NAVIGATION US260 FIXED-SATELLITE (Earth-to-space) 5.447A US344	RF Devices (15) *Satellite Communications* (25) Aviation (87)
5.446 5.446C 5.447 5.447B 5.447C			US211 US307 US344	5.447C US211 US307	

(*Continued*)

TABLE F.1 (*Continued*)

Worldwide Frequency Allocations

Table of Frequency Allocations   4,990–5,925 MHz (SHF)

International Table			United States Table		
Region 1 Table	Region 2 Table	Region 3 Table	Federal Table	Non-Federal Table	FCC Rule Part(s)
5250–5255 EARTH EXPLORATION-SATELLITE (active) RADIOLOCATION SPACE RESEARCH 5.447D MOBILE except aeronautical mobile 5.446A 5.447F			5250–5255 EARTH EXPLORATION-SATELLITE (active) RADIOLOCATION G59 SPACE RESEARCH (active) 5.447D	5250–5255 Earth exploration-satellite (active) Radiolocation Space research	RF Devices (15) Private Land Mobile (90)
5.447E 5.448 5.448A			5.448A	5.448A	
5255–5350 EARTH EXPLORATION-SATELLITE (active) RADIOLOCATION SPACE RESEARCH (active) MOBILE except aeronautical mobile 5.446A 5.447F			5255–5350 EARTH EXPLORATION-SATELLITE (active) RADIOLOCATION G59 SPACE RESEARCH (active)	5255–5350 Earth exploration-satellite (active) Radiolocation Space research (active)	
5.447E 5.448 5.448A			5.448A	5.448A	

(Continued)

TABLE F.1 (*Continued*)

Worldwide Frequency Allocations

Table of Frequency Allocations				4,990–5,925 MHz (SHF)		
International Table				United States Table		
Region 1 Table	Region 2 Table	Region 3 Table		Federal Table	Non-Federal Table	FCC Rule Part(s)
5350–5460 EARTH EXPLORATION-SATELLITE (active) 5.448B SPACE RESEARCH (active) 5.448C AERONAUTICAL RADIONAVIGATION 5.449 RADIOLOCATION 5.448D				5350–5460 EARTH EXPLORATION-SATELLITE (active) 5.448B SPACE RESEARCH (active) AERONAUTICAL RADIO-NAVIGATION 5.449 RADIOLOCATION G56	5350–5460 AERONAUTICAL RADIO-NAVIGATION 5.449 Earth exploration-satellite (active) 5.448B Space research (active) Radiolocation	Aviation (87) Private Land Mobile (90)
				US390 G130	US390	

(Continued)

TABLE F.1 (*Continued*)

Worldwide Frequency Allocations

Table of Frequency Allocations | 4,990–5,925 MHz (SHF)

International Table			United States Table		FCC Rule Part(s)
Region 1 Table	Region 2 Table	Region 3 Table	Federal Table	Non-Federal Table	
5460–5470 RADIONAVIGATION 5.449 EARTH EXPLORATION-SATELLITE (active) SPACE RESEARCH (active) RADIOLOCATION 5.448D			5460–5470 RADIO-NAVIGATION 5.449 US65 EARTH EXPLORATION-SATELLITE (active) SPACE RESEARCH (active) RADIOLOCATION G56	5460–5470 RADIO-NAVIGATION 5.449 US65 Earth exploration-satellite (active) Space research (active) Radiolocation	Maritime (80) Aviation (87) Private Land Mobile (90)
5.448B			5.448B US49 G130	5.448B US49	

(Continued)

TABLE F.1 (*Continued*)

Worldwide Frequency Allocations

Table of Frequency Allocations			4,990–5,925 MHz (SHF)		
International Table			United States Table		
Region 1 Table	Region 2 Table	Region 3 Table	Federal Table	Non-Federal Table	FCC Rule Part(s)
5470–5570 MARITIME RADIONAVIGATION MOBILE except aeronautical mobile 5.446A 5.450A EARTH EXPLORATION-SATELLITE (active) SPACE RESEARCH (active) RADIOLOCATION 5.450B			5470–5570 MARITIME RADIO-NAVIGATION US65 EARTH EXPLORATION-SATELLITE (active) SPACE RESEARCH (active) RADIOLOCATION G56	5470–5570 MARITIME RADIO-NAVIGATION US65 RADIOLOCATION Earth exploration-satellite (active) Space research (active)	RF Devices (15) Maritime (80) Private Land Mobile (90)
5.448B 5.450 5.451			5.448B US50 G131	US50	
5570–5650 MARITIME RADIONAVIGATION MOBILE except aeronautical mobile 5.446A 5.450A RADIOLOCATION 5.450B			5570–5600 MARITIME RADIO-NAVIGATION US65 RADIOLOCATION G56	5570–5600 MARITIME RADIO-NAVIGATION US65 RADIOLOCATION	
			US50 G131	US50	

(Continued)

TABLE F.1 (Continued)
Worldwide Frequency Allocations

Table of Frequency Allocations — 4,990–5,925 MHz (SHF)

International Table			United States Table		FCC Rule Part(s)
Region 1 Table	Region 2 Table	Region 3 Table	Federal Table	Non-Federal Table	
5.450 5.451 5.452			5600–5650 MARITIME RADIO-NAVIGATION US65 METEOROLOGICAL AIDS RADIOLOCATION G56	5600–5650 MARITIME RADIO-NAVIGATION US65 METEOROLOGICAL AIDS RADIOLOCATION	
5650–5725 MOBILE except aeronautical mobile 5.446A 5.450A RADIOLOCATION Amateur Space research (deep space)			5.452 US50 G131	5.452 US50	RF Devices (15) ISM Equipment (18) Amateur Radio (97)
			5650–5925 RADIOLOCATION G2	5650–5830 Amateur	
5.282 5.451 5.453 5.454 5.455					
5725–5830 FIXED-SATELLITE (Earth-to-space) RADIOLOCATION Amateur 5.456	5725–5830 RADIOLOCATION Amateur				
5.150 5.451 5.453 5.455	5.150 5.453 5.455		5.150 5.282	5.150 5.282	

(Continued)

TABLE F.1 (Continued)

Worldwide Frequency Allocations

Table of Frequency Allocations   4,990–5,925 MHz (SHF)

International Table			United States Table		FCC Rule Part(s)
Region 1 Table	Region 2 Table	Region 3 Table	Federal Table	Non-Federal Table	
5830–5850 FIXED-SATELLITE (Earth-to-space) RADIOLOCATION Amateur Amateur-satellite (space-to-Earth)	5830–5850 RADIOLOCATION Amateur Amateur-satellite (space-to-Earth)			5830–5850 Amateur Amateur-satellite (space-to-Earth)	
5.150 5.451 5.453 5.455 5.456	5.150 5.453 5.455			5.150	
5850–5925 FIXED FIXED-SATELLITE (Earth-to-space) MOBILE	5850–5925 FIXED FIXED-SATELLITE (Earth-to-space) MOBILE Amateur Radiolocation	5850–5925 FIXED FIXED-SATELLITE (Earth-to-space) MOBILE Radiolocation		5850–5925 FIXED-SATELLITE (Earth-to-space) US245 MOBILE NG160 Amateur	ISM Equipment (18) Private Land Mobile (90) Personal Radio (95) Amateur Radio (97)
5.150	5.150	5.150	5.150 US245	5.150	

(Continued)

TABLE F.1 (*Continued*)

Worldwide Frequency Allocations

Table of Frequency Allocations   5,925–8,025 MHz (SHF)

International Table			United States Table		FCC Rule Part(s)
Region 1 Table	Region 2 Table	Region 3 Table	Federal Table	Non-Federal Table	
5925–6700 FIXED FIXED-SATELLITE (Earth-to-space) 5.457A 5.457B MOBILE 5.457C			5925–6425	5925–6425 FIXED FIXED-SATELLITE (Earth-to-space) NG181	Satellite Communications (25) Fixed Microwave (101)
			6425–6525	6425–6525 FIXED-SATELLITE (Earth-to-space) MOBILE	TV Broadcast Auxiliary (74F) Cable TV Relay (78) Fixed Microwave (101)
			5.440 5.458	5.440 5.458	
			6525–6700	6525–6700 FIXED FIXED-SATELLITE (Earth-to-space)	Fixed Microwave (101)
5.149 5.440 5.458			5.458 US342	5.458 US342	
6700–7075 FIXED FIXED-SATELLITE (Earth-to-space) (space-to-Earth) 5.441 MOBILE			6700–7125	6700–6875 FIXED FIXED-SATELLITE (Earth-to-space) (space-to-Earth) 5.441	Satellite Communications (25) Fixed Microwave (101)
				5.458 5.458A 5.458B	

(*Continued*)

TABLE F.1 (*Continued*)

Worldwide Frequency Allocations

Table of Frequency Allocations   5,925–8,025 MHz (SHF)

International Table			United States Table		FCC Rule Part(s)
Region 1 Table	Region 2 Table	Region 3 Table	Federal Table	Non-Federal Table	
5.458 5.458A 5.458B 5.458C				6875–7025 FIXED NG118 FIXED-SATELLITE (Earth-to-space) (space-to-Earth) 5.441 MOBILE NG171 5.458 5.458A 5.458B	Satellite Communications (25) TV Broadcast Auxiliary (74F) Cable TV Relay (78)
				7025–7075 FIXED NG118 FIXED-SATELLITE (Earth-to-space) NG172 MOBILE NG171 5.458 5.458A 5.458B	TV Broadcast Auxiliary (74F) Cable TV Relay (78)
7075–7145 FIXED MOBILE			5.458	7075–7125 FIXED NG118 MOBILE NG171 5.458	

(*Continued*)

TABLE F.1 (*Continued*)

Worldwide Frequency Allocations

Table of Frequency Allocations | 5,925–8,025 MHz (SHF)

International Table			United States Table		FCC Rule Part(s)
Region 1 Table	Region 2 Table	Region 3 Table	Federal Table	Non-Federal Table	
5.458 5.459			7125–7145 FIXED 5.458 G116	7125–7235	
7145–7235 FIXED MOBILE SPACE RESEARCH (Earth-to-space) 5.460			7145–7190 FIXED SPACE RESEARCH (deep space) (Earth-to-space) US262 5.458 G116		
			7190–7235 FIXED SPACE RESEARCH (Earth-to-space) G133		
5.458 5.459			5.458 G134	5.458 US262	

(Continued)

TABLE F.1 (*Continued*)

Worldwide Frequency Allocations

Table of Frequency Allocations

5,925–8,025 MHz (SHF)

International Table			United States Table		FCC Rule Part(s)
Region 1 Table	Region 2 Table	Region 3 Table	Federal Table	Non-Federal Table	
7235–7250 FIXED MOBILE			7235–7250 FIXED	7235–7250	
5.458			5.458	5.458	
7250–7300 FIXED FIXED-SATELLITE (space-to-Earth) MOBILE			7250–7300 FIXED-SATELLITE (space-to-Earth) MOBILE-SATELLITE (space-to-Earth) Fixed	7250–8025	
5.461			G117		
7300–7450 FIXED FIXED-SATELLITE (space-to-Earth) MOBILE except aeronautical mobile			7300–7450 FIXED FIXED-SATELLITE (space-to-Earth) Mobile-satellite (space-to-Earth)		
5.461			G117		

(Continued)

TABLE F.1 (*Continued*)

Worldwide Frequency Allocations

Table of Frequency Allocations			5,925–8,025 MHz (SHF)		
International Table			**United States Table**		
Region 1 Table	Region 2 Table	Region 3 Table	Federal Table	Non-Federal Table	FCC Rule Part(s)
7450–7550 FIXED FIXED-SATELLITE (space-to-Earth) METEOROLOGICAL-SATELLITE (space-to-Earth) MOBILE except aeronautical mobile			7450–7550 FIXED FIXED-SATELLITE (space-to-Earth) METEORO- LOGICAL- SATELLITE (space-to-Earth) Mobile-satellite (space-to-Earth) G104 G117		
5.461A					
7550–7750 FIXED FIXED-SATELLITE (space-to-Earth) MOBILE except aeronautical mobile			7550–7750 FIXED FIXED-SATELLITE (space-to-Earth) Mobile-satellite (space-to-Earth) G117		

(Continued)

TABLE F.1 (*Continued*)

Worldwide Frequency Allocations

Table of Frequency Allocations

5,925–8,025 MHz (SHF)

International Table			United States Table		FCC Rule Part(s)
Region 1 Table	Region 2 Table	Region 3 Table	Federal Table	Non-Federal Table	
7750–7850 FIXED METEOROLOGICAL-SATELLITE (space-to-Earth) 5.461B MOBILE except aeronautical mobile			7750–7850 FIXED METEORO- LOGICAL- SATELLITE (space-to-Earth) 5.461B		
7850–7900 FIXED MOBILE except aeronautical mobile			7850–7900 FIXED		
7900–8025 FIXED FIXED-SATELLITE (Earth-to-space) MOBILE 5.461			7900–8025 FIXED-SATELLITE (Earth-to-space) MOBILE-SATELLITE (Earth-to-space) Fixed G117		

(*Continued*)

TABLE F.1 (*Continued*)

Worldwide Frequency Allocations

Table of Frequency Allocations

8,025–10,000 MHz (SHF)

International Table			United States Table		
Region 1 Table	Region 2 Table	Region 3 Table	Federal Table	Non-Federal Table	FCC Rule Part(s)
8025–8175 EARTH EXPLORATION-SATELLITE (space-to-Earth) FIXED FIXED-SATELLITE (Earth-to-space) MOBILE 5.463			8025–8175 EARTH EXPLORATION- SATELLITE (space-to-Earth) FIXED FIXED-SATELLITE (Earth-to-space) Mobile-satellite (Earth-to-space) (no airborne transmissions)	8025–8400	
5.462A			US258 G117		

(*Continued*)

TABLE F.1 (*Continued*)
Worldwide Frequency Allocations

Table of Frequency Allocations · 8,025–10,000 MHz (SHF)

International Table			United States Table		
Region 1 Table	Region 2 Table	Region 3 Table	Federal Table	Non-Federal Table	FCC Rule Part(s)
8175–8215 EARTH EXPLORATION-SATELLITE (space-to-Earth) FIXED FIXED-SATELLITE (Earth-to-space) METEOROLOGICAL-SATELLITE (Earth-to-space) MOBILE 5.463 5.462A			8175–8215 EARTH EXPLORATION-SATELLITE (space-to-Earth) FIXED FIXED-SATELLITE (Earth-to-space) METEORO-LOGICAL-SATELLITE (Earth-to-space) Mobile-satellite (Earth-to-space) (no airborne transmissions) US258 G104 G117		

(Continued)

TABLE F.1 (*Continued*)

Worldwide Frequency Allocations

Table of Frequency Allocations   8,025–10,000 MHz (SHF)

International Table			United States Table		FCC Rule Part(s)
Region 1 Table	Region 2 Table	Region 3 Table	Federal Table	Non-Federal Table	
8215–8400 EARTH EXPLORATION-SATELLITE (space-to-Earth) FIXED FIXED-SATELLITE (Earth-to-space) MOBILE 5.463			8215–8400 EARTH EXPLORATION-SATELLITE (space-to-Earth) FIXED FIXED-SATELLITE (Earth-to-space) Mobile-satellite (Earth-to-space) (no airborne transmissions)		
5.462A			US258 G117	US258	
8400–8500 FIXED MOBILE except aeronautical mobile SPACE RESEARCH (space-to-Earth) 5.465 5.466			8400–8450 FIXED SPACE RESEARCH (deep space) (space-to-Earth)	8400–8450 Space research (deep space) (space-to-Earth)	
			8450–8500 FIXED SPACE RESEARCH (space-to-Earth)	8450–8500 SPACE RESEARCH (space-to-Earth)	

(*Continued*)

TABLE F.1 (Continued)

Worldwide Frequency Allocations

Table of Frequency Allocations — 8,025–10,000 MHz (SHF)

International Table			United States Table		FCC Rule Part(s)
Region 1 Table	Region 2 Table	Region 3 Table	Federal Table	Non-Federal Table	
8500–8550 RADIOLOCATION 5.468 5.469			8500–8550 RADIOLOCATION G59	8500–8550 Radiolocation	Private Land Mobile (90)
8550–8650 EARTH EXPLORATION-SATELLITE (active) RADIOLOCATION SPACE RESEARCH (active) 5.468 5.469 5.469A			8550–8650 EARTH EXPLORATION-SATELLITE (active) RADIOLOCATION G59 SPACE RESEARCH (active)	8550–8650 Earth exploration-satellite (active) Radiolocation Space research (active)	
8650–8750 RADIOLOCATION 5.468 5.469			8650–9000 RADIOLOCATION G59	8650–9000 Radiolocation	Aviation (87) Private Land Mobile (90)
8750–8850 RADIOLOCATION AERONAUTICAL RADIONAVIGATION 5.470 5.471					

(Continued)

TABLE F.1 (*Continued*)

Worldwide Frequency Allocations

Table of Frequency Allocations   8,025–10,000 MHz (SHF)

International Table			United States Table			FCC Rule Part(s)
Region 1 Table	Region 2 Table	Region 3 Table	Federal Table	Non-Federal Table		
8850–9000 RADIOLOCATION MARITIME RADIONAVIGATION 5.472						
5.473			US53	US53		
9000–9200 AERONAUTICAL RADIONAVIGATION 5.337 RADIOLOCATION	AERONAUTICAL RADIONAVIGATION 5.337		9000–9200 AERONAUTICAL RADIO- NAVIGATION 5.337 Radiolocation G2	9000–9200 AERONAUTICAL RADIO- NAVIGATION 5.337 Radiolocation		
5.471 5.473A			US48 G19	US48		
9200–9300 RADIOLOCATION MARITIME RADIONAVIGATION 5.472	MARITIME RADIONAVIGATION 5.472		9200–9300 MARITIME RADIO- NAVIGATION 5.472 Radiolocation US110 G59	9200–9300 MARITIME RADIO- NAVIGATION 5.472 Radiolocation US110		Maritime (80) Private Land Mobile (90)
5.473 5.474			5.474	5.474		

(*Continued*)

TABLE F.1 (*Continued*)

Worldwide Frequency Allocations

Table of Frequency Allocations — 8,025–10,000 MHz (SHF)

International Table			United States Table		FCC Rule Part(s)
Region 1 Table	Region 2 Table	Region 3 Table	Federal Table	Non-Federal Table	
9300–9500 EARTH EXPLORATION-SATELLITE (active) SPACE RESEARCH (active) RADIOLOCATION RADIONAVIGATION 5.427 5.474 5.475 5.475A 5.475B 5.476A			9300–9500 RADIO-NAVIGATION US66 Radiolocation US51 G56 Meteorological aids 5.427 5.474 US67 US71	9300–9500 RADIO-NAVIGATION US66 Radiolocation US51 Meteorological aids 5.427 5.474 US67 US71	Maritime (80) Aviation (87) Private Land Mobile (90)
9500–9800 EARTH EXPLORATION-SATELLITE (active) SPACE RESEARCH (active) RADIOLOCATION RADIONAVIGATION 5.476A			9500–9800 EARTH EXPLORATION-SATELLITE (active) SPACE RESEARCH (active) RADIOLOCATION	9500–9800 Earth exploration-satellite (active) Space research (active) Radiolocation	Private Land Mobile (90)
9800–9900 RADIOLOCATION Earth exploration-satellite (active) Space research (active) Fixed 5.477 5.478 5.478A 5.478B			9800–10000 RADIOLOCATION 5.479	9800–10000 Radiolocation 5.479	

(*Continued*)

TABLE F.1 (*Continued*)

Worldwide Frequency Allocations

Table of Frequency Allocations			8,025–10,000 MHz (SHF)		
International Table			United States Table		
Region 1 Table	Region 2 Table	Region 3 Table	Federal Table	Non-Federal Table	FCC Rule Part(s)
9900–10000 RADIOLOCATION Fixed					
5.477 5.478 5.479					

Table of Frequency Allocations | 10–14 GHz (SHF)

Region 1 Table	Region 2 Table	Region 3 Table	Federal Table	Non-Federal Table	FCC Rule Part(s)
10–10.45 FIXED MOBILE RADIOLOCATION Amateur	10–10.45 RADIOLOCATION Amateur	10–10.45 FIXED MOBILE RADIOLOCATION Amateur	10–10.5 RADIOLOCATION US108 G32	10–10.45 Amateur Radiolocation US108	Private Land Mobile (90) Amateur Radio (97)
5.479	5.479 5.480	5.479		5.479 US128 NG50	
10.45–10.5 RADIOLOCATION Amateur Amateur-satellite				10.45–10.5 Amateur Amateur-satellite Radiolocation US108	
5.481			5.479 US128	US128 NG50	
10.5–10.55 FIXED MOBILE RADIOLOCATION Radiolocation	10.5–10.55 FIXED MOBILE RADIOLOCATION		10.5–10.55 RADIOLOCATION US59		Private Land Mobile (90)

(*Continued*)

TABLE F.1 (*Continued*)

Worldwide Frequency Allocations

Table of Frequency Allocations

10–14 GHz (SHF)

International Table			United States Table		FCC Rule Part(s)
Region 1 Table	Region 2 Table	Region 3 Table	Federal Table	Non-Federal Table	
10.55–10.6 FIXED MOBILE except aeronautical mobile Radiolocation			10.55–10.6	10.55–10.6 FIXED	Fixed Microwave (101)
10.6–10.68 EARTH EXPLORATION-SATELLITE (passive) FIXED MOBILE except aeronautical mobile RADIO ASTRONOMY SPACE RESEARCH (passive) Radiolocation			10.6–10.68 EARTHEXPLORATION-SATELLITE (passive) SPACE RESEARCH (passive)	10.6–10.68 EARTHEXPLORATION-SATELLITE (passive) FIXED US265 SPACE RESEARCH (passive)	
5.149 5.482 5.482A			US130 US131 US265	US130 US131	
10.68–10.7 EARTH EXPLORATION-SATELLITE (passive) RADIO ASTRONOMY SPACE RESEARCH (passive)			10.68–10.7 EARTH EXPLORATION-SATELLITE (passive) RADIO ASTRONOMY US74 SPACE RESEARCH (passive)		
5.340 5.483			US131 US246		

(Continued)

TABLE F.1 (*Continued*)

Worldwide Frequency Allocations

Table of Frequency Allocations

10–14 GHz (SHF)

International Table			United States Table		FCC Rule Part(s)
Region 1 Table	Region 2 Table	Region 3 Table	Federal Table	Non-Federal Table	
10.7–11.7 FIXED FIXED-SATELLITE (space-to-Earth) 5.441 5.484A (Earth-to-space) 5.484 MOBILE except aeronautical mobile	10.7–11.7 FIXED FIXED-SATELLITE (space-to-Earth) 5.484A MOBILE except aeronautical mobile	10.7–11.7 FIXED FIXED-SATELLITE (space-to-Earth) 5.441 5.484A MOBILE except aeronautical mobile	10.7–11.7	10.7–11.7 FIXED FIXED-SATELLITE (space-to-Earth) 5.441 US131 US211 NG52	Satellite Communications (25) Fixed Microwave (101)
			US131 US211		
11.7–12.5 FIXED MOBILE except aeronautical mobile BROADCASTING BROADCASTING-SATELLITE 5.492	11.7–12.1 FIXED 5.486 FIXED-SATELLITE (space-to-Earth) 5.484A 5.488 Mobile except aeronautical mobile 5.485	11.7–12.2 FIXED MOBILE except aeronautical mobile BROADCASTING BROADCASTING-SATELLITE 5.492	11.7–12.2	11.7–12.2 FIXED-SATELLITE (space-to-Earth) 5.485 5.488 NG55 NG143	Satellite Communications (25)
	12.1–12.2 FIXED-SATELLITE (space-to-Earth) 5.484A 5.488 5.485 5.489	5.487 5.487A			

(*Continued*)

TABLE F.1 (*Continued*)

Worldwide Frequency Allocations

Table of Frequency Allocations

10–14 GHz (SHF)

International Table			United States Table		FCC Rule Part(s)
Region 1 Table	Region 2 Table	Region 3 Table	Federal Table	Non-Federal Table	
	12.2–12.7 FIXED MOBILE except aeronautical mobile BROADCASTING BROADCASTING-SATELLITE 5.492	12.2–12.5 FIXED FIXED-SATELLITE (space-to-Earth) MOBILE except aeronautical mobile BROADCASTING	12.2–12.75	12.2–12.7 FIXED BROADCASTING-SATELLITE	Satellite Communications (25) Fixed Microwave (101)
5.487 5.487A		5.484A 5.487			
12.5–12.75 FIXED-SATELLITE (space-to-Earth) 5.484A (Earth-to-space)	5.487A 5.488 5.490 12.7–12.75 FIXED FIXED-SATELLITE (Earth-to-space) MOBILE except aeronautical mobile	12.5–12.75 FIXED FIXED-SATELLITE (space-to-Earth) 5.484A MOBILE except aeronautical mobile BROADCASTING-SATELLITE 5.493		5.487A 5.488 5.490 12.7–12.75 FIXED NG118 FIXED-SATELLITE (Earth-to-space) MOBILE	TV Broadcast Auxiliary (74F) Cable TV Relay (78) Fixed Microwave (101)

(*Continued*)

TABLE F.1 (*Continued*)

Worldwide Frequency Allocations

Table of Frequency Allocations | 10–14 GHz (SHF)

International Table			United States Table		FCC Rule Part(s)
Region 1 Table	Region 2 Table	Region 3 Table	Federal Table	Non-Federal Table	
12.75–13.25 FIXED FIXED-SATELLITE (Earth-to-space) 5.441 MOBILE Space research (deep space) (space-to-Earth)			12.75–13.25	12.75–13.25 FIXED NG118 FIXED-SATELLITE (Earth-to-space) 5.441 NG52 MOBILE	Satellite Communications (25) TV Broadcast Auxiliary (74F) Cable TV Relay (78) Fixed Microwave (101)
			US251	US251 NG53	
13.25–13.4 EARTH EXPLORATION-SATELLITE (active) AERONAUTICAL RADIONAVIGATION 5.497 SPACE RESEARCH (active)			13.25–13.4 EARTH EXPLORATION-SATELLITE (active) AERONAUTICAL RADIO-NAVIGATION 5.497 SPACE RESEARCH (active)	13.25–13.4 AERONAUTICAL RADIO-NAVIGATION 5.497 Earth exploration-satellite (active) Space research (active)	Aviation (87)
5.498A 5.499			5.498A		

(*Continued*)

TABLE F.1 (*Continued*)

Worldwide Frequency Allocations

Table of Frequency Allocations

10–14 GHz (SHF)

International Table			United States Table		FCC Rule Part(s)
Region 1 Table	Region 2 Table	Region 3 Table	Federal Table	Non-Federal Table	
13.4–13.75 EARTH EXPLORATION-SATELLITE (active) RADIOLOCATION SPACE RESEARCH 5.501A Standard frequency and time signal-satellite (Earth-to-space) 5.499 5.500 5.501 5.501B			13.4–13.75 EARTH EXPLORATION-SATELLITE (active) RADIOLOCATION G59 SPACE RESEARCH 5.501A Standard frequency and time signal-satellite (Earth-to-space) 5.501B	13.4–13.75 Earth exploration-satellite (active) Radiolocation Space research Standard frequency and time signal-satellite (Earth-to-space)	Private Land Mobile (90)
13.75–14 FIXED-SATELLITE (Earth-to-space) 5.484A RADIOLOCATION Earth exploration-satellite Standard frequency and time signal-satellite (Earth-to-space) Space research 5.499 5.500 5.501 5.502 5.503			13.75–14 RADIOLOCATION G59 Standard frequency and time signal-satellite (Earth-to-space) Space research US337 US356 US357	13.75–14 FIXED-SATELLITE (Earth-to-space) US337 Standard frequency and time signal-satellite (Earth-to-space) Space research Radiolocation US356 US357	Satellite Communications (25) Private Land Mobile (90)

(Continued)

TABLE F.1 (*Continued*)

Worldwide Frequency Allocations

Table of Frequency Allocations | | | 14–17.7 GHz (SHF) | | |

International Table			United States Table		FCC Rule Part(s)
Region 1 Table	Region 2 Table	Region 3 Table	Federal Table	Non-Federal Table	
14–14.25 FIXED-SATELLITE (Earth-to-space) 5.457A 5.457B 5.484A 5.506 5.506B RADIONAVIGATION 5.504 Mobile-satellite (Earth-to-space) 5.504B 5.504C 5.506A Space research 5.504A 5.505			14–14.2 Space research US133	14–14.2 FIXED-SATELLITE (Earth-to-space) NG55 Mobile-satellite (Earth-to-space) Space research US133	Satellite Communications (25)
14.25–14.3 FIXED-SATELLITE (Earth-to-space) 5.457A 5.457B 5.484A 5.506 5.506B RADIONAVIGATION 5.504 Mobile-satellite (Earth-to-space) 5.504B 5.506A 5.508A Space research 5.504A 5.505 5.508			14.2–14.4	14.2–14.47 FIXED-SATELLITE (Earth-to-space) NG55 Mobile-satellite (Earth-to-space)	

(*Continued*)

TABLE F.1 (*Continued*)

Worldwide Frequency Allocations

Table of Frequency Allocations

	14–17.7 GHz (SHF)				
International Table			**United States Table**		**FCC Rule Part(s)**
Region 1 Table	**Region 2 Table**	**Region 3 Table**	**Federal Table**	**Non-Federal Table**	
14.3–14.4 FIXED FIXED-SATELLITE (Earth-to-space) 5.457A 5.457B 5.484A 5.506 5.506B MOBILE except aeronautical mobile Mobile-satellite (Earth-to-space) 5.506A 5.509A Radionavigation- satellite	14.3–14.4 FIXED-SATELLITE (Earth-to-space) 5.457A 5.484A 5.506 5.506B Mobile-satellite (Earth-to-space) 5.506A Radionavigation- satellite	14.3–14.4 FIXED FIXED-SATELLITE (Earth-to-space) 5.457A 5.484A 5.506 5.506B MOBILE except aeronautical mobile Mobile-satellite (Earth-to-space) 5.504B 5.506A 5.509A Radionavigation- satellite			
5.504A	5.504A	5.504A			
14.4–14.47 FIXED FIXED-SATELLITE (Earth-to-space) 5.457A 5.457B 5.484A 5.506 5.506B MOBILE except aeronautical mobile Mobile-satellite (Earth-to-space) 5.504B 5.506A 5.509A Space research (space-to-Earth)			14.4–14.47 Fixed Mobile		
5.504A					

(Continued)

TABLE F.1 (Continued)

Worldwide Frequency Allocations

Table of Frequency Allocations

International Table			United States Table		FCC Rule Part(s)
			14–17.7 GHz (SHF)		
Region 1 Table	Region 2 Table	Region 3 Table	Federal Table	Non-Federal Table	
14.47–14.5 FIXED FIXED-SATELLITE (Earth-to-space) 5.457A 5.457B 5.484A 5.506 5.506B MOBILE except aeronautical mobile Mobile-satellite (Earth-to-space) 5.504B 5.506A 5.509A Radio astronomy			14.47–14.5 Fixed Mobile	14.47–14.5 FIXED-SATELLITE (Earth-to-space) NG55 Mobile-satellite (Earth-to-space)	
5.149 5.504A			US133 US203 US342	US133 US203 US342	
14.5–14.8 FIXED FIXED-SATELLITE (Earth-to-space) 5.510 MOBILE Space research			14.5–14.7145 FIXED Mobile Space research	14.5–14.8	
			14.7145–14.8 MOBILE Fixed Space research		
14.8–15.35 FIXED MOBILE Space research			14.8–15.1365 MOBILE SPACE RESEARCH Fixed	14.8–15.1365	
			US310	US310	

(Continued)

TABLE F.1 (*Continued*)

Worldwide Frequency Allocations

Table of Frequency Allocations

14–17.7 GHz (SHF)

International Table			United States Table		FCC Rule Part(s)
Region 1 Table	Region 2 Table	Region 3 Table	Federal Table	Non-Federal Table	
5.339			15.1365–15.35 FIXED SPACE RESEARCH Mobile	15.1365–15.35	
			5.339 US211	5.339 US211	
15.35–15.4 EARTH EXPLORATION-SATELLITE (passive) RADIO ASTRONOMY SPACE RESEARCH (passive)			15.35–15.4 EARTH EXPLORATION-SATELLITE (passive) RADIO ASTRONOMY US74 SPACE RESEARCH (passive)		
5.340 5.511			US246		
15.4–15.43 AERONAUTICAL RADIONAVIGATION			15.4–15.43 AERONAUTICAL RADIONAVIGATION US260		Aviation (87)
5.511D			US211		

(Continued)

TABLE F.1 (*Continued*)

Worldwide Frequency Allocations

Table of Frequency Allocations

14–17.7 GHz (SHF)

International Table			United States Table		FCC Rule Part(s)
Region 1 Table	Region 2 Table	Region 3 Table	Federal Table	Non-Federal Table	
15.43–15.63 FIXED-SATELLITE (Earth-to-space) 5.511A AERONAUTICAL RADIONAVIGATION			15.43–15.63 AERONAUTICAL RADIO-NAVIGATION US260	15.43–15.63 FIXED-SATELLITE (Earth-to-space) AERONAUTICAL RADIO-NAVIGATION US260	Satellite Communications (25) Aviation (87)
5.511C			5.511C US211 US359	5.511C US211 US359	
15.63–15.7 AERONAUTICAL RADIONAVIGATION			15.63–15.7 AERONAUTICAL RADIONAVIGATION US260	15.7–17.2 Radiolocation	Aviation (87)
5.511D			US211		
15.7–16.6 RADIOLOCATION 5.512 5.513			15.7–16.6 RADIOLOCATION G59		Private Land Mobile (90)
16.6–17.1 RADIOLOCATION Space research (deep space) (Earth-to-space) 5.512 5.513			16.6–17.1 RADIOLOCATION G59 Space research (deep space) (Earth-to-space)		

(*Continued*)

TABLE F.1 (*Continued*)

Worldwide Frequency Allocations

Table of Frequency Allocations

14–17.7 GHz (SHF)

International Table			United States Table		FCC Rule Part(s)
Region 1 Table	Region 2 Table	Region 3 Table	Federal Table	Non-Federal Table	
17.1–17.2 RADIOLOCATION 5.512 5.513			17.1–17.2 RADIOLOCATION G59		
17.2–17.3 EARTH EXPLORATION-SATELLITE (active) RADIOLOCATION SPACE RESEARCH (active) 5.512 5.513 5.513A			17.2–17.3 EARTH EXPLORATION-SATELLITE (active) RADIOLOCATION G59 SPACE RESEARCH (active)	17.2–17.3 Earth exploration-satellite (active) Radiolocation Space research (active)	
17.3–17.7 FIXED-SATELLITE (Earth-to-space) 5.516 (space-to-Earth) 5.516A 5.516B Radiolocation 5.514	17.3–17.7 FIXED-SATELLITE (Earth-to-space) 5.516 BROADCASTING-SATELLITE Radiolocation 5.514 5.515	17.3–17.7 FIXED-SATELLITE (Earth-to-space) 5.516 Radiolocation 5.514	17.3–17.7 Radiolocation US259 G59 US402 G117	17.3–17.7 FIXED-SATELLITE (Earth-to-space) US271 BROADCASTING-SATELLITE US402 NG163 US259	Satellite Communications (25)

(*Continued*)

TABLE F.1 (Continued)

Worldwide Frequency Allocations

Table of Frequency Allocations   **17.7–23.6 GHz (SHF)**

International Table			United States Table		FCC Rule Part(s)
Region 1 Table	Region 2 Table	Region 3 Table	Federal Table	Non-Federal Table	
17.7–18.1 FIXED FIXED-SATELLITE (space-to-Earth) 5.484A (Earth-to-space) 5.516 MOBILE	17.7–17.8 FIXED FIXED-SATELLITE (space-to-Earth) 5.517 (Earth-to-space) 5.516 BROADCASTING-SATELLITE Mobile	17.7–18.1 FIXED FIXED-SATELLITE (space-to-Earth) 5.484A (Earth-to-space) 5.516 MOBILE	17.7–17.8	17.7–17.8 FIXED NG144 FIXED-SATELLITE (Earth-to-space) US271	Satellite Communications (25) TV Broadcast Auxiliary (74F) Cable TV Relay (78) Fixed Microwave (101)
	5.515		US401 G117	US401	
	17.8–18.1 FIXED FIXED-SATELLITE (space-to-Earth) 5.484A (Earth-to-space) 5.516 MOBILE 5.519		17.8–18.3 FIXED-SATELLITE (space-to-Earth) US334 G117	17.8–18.3 FIXED NG144	TV Broadcast Auxiliary (74F) Cable TV Relay (78) Fixed Microwave (101)

(Continued)

TABLE F.1 (*Continued*)

Worldwide Frequency Allocations

Table of Frequency Allocations — 17.7–23.6 GHz (SHF)

International Table			United States Table		FCC Rule Part(s)
Region 1 Table	Region 2 Table	Region 3 Table	Federal Table	Non-Federal Table	
18.1–18.4 FIXED FIXED-SATELLITE (space-to-Earth) 5.484A 5.516B (Earth-to-space) 5.520 MOBILE 5.519 5.521			US519	US334 US519	
18.4–18.6 FIXED FIXED-SATELLITE (space-to-Earth) 5.484A 5.516B MOBILE			18.3–18.6 FIXED-SATELLITE (space-to-Earth) US334 G117	18.3–18.6 FIXED-SATELLITE (space-to-Earth) NG164	Satellite Communications (25)
				US334 NG144	
18.6–18.8 EARTH EXPLORATION-SATEL-LITE (passive) FIXED FIXED-SATELLITE (space-to-Earth) 5.522B MOBILE except aeronautical mobile Space research (passive)	18.6–18.8 EARTH EXPLORATION-SATELLITE (passive) FIXED FIXED-SATELLITE (space-to-Earth) 5.516B 5.522B MOBILE except aeronautical mobile SPACE RESEARCH (passive)	18.6–18.8 EARTH EXPLORATION-SATELLITE (passive) FIXED FIXED-SATELLITE (space-to-Earth) 5.522B MOBILE except aeronautical mobile Space research (passive)	18.6–18.8 EARTH EXPLORATION-SATELLITE (passive) FIXED-SATELLITE (space-to-Earth) US255 US334 G117 SPACE RESEARCH (passive)	18.6–18.8 EARTH EXPLORATION-SATELLITE (passive) FIXED-SATELLITE (space-to-Earth) US255 NG164 SPACE RESEARCH (passive)	
5.522A 5.522C	5.522A	5.522A	US254	US254 US334 NG144	

(Continued)

TABLE F.1 (Continued)

Worldwide Frequency Allocations

Table of Frequency Allocations   17.7–23.6 GHz (SHF)

International Table			United States Table		FCC Rule Part(s)
Region 1 Table	Region 2 Table	Region 3 Table	Federal Table	Non-Federal Table	
18.8–19.3 FIXED FIXED-SATELLITE (space-to-Earth) 5.516B 5.523A MOBILE	FIXED-SATELLITE (space-to-Earth) 5.516B 5.523A		18.8–20.2 FIXED-SATELLITE (space-to-Earth) US334 G117	18.8–19.3 FIXED-SATELLITE (space-to-Earth) NG165 US334 NG144	
19.3–19.7 FIXED FIXED-SATELLITE (space-to-Earth) (Earth-to-space) 5.523E MOBILE	FIXED-SATELLITE (space-to-Earth) (Earth-to-space) 5.523B 5.523C 5.523D			19.3–19.7 FIXED NG144 FIXED-SATELLITE (space-to-Earth) NG166 US334	Satellite Communications (25) TV Broadcast Auxiliary (74F) Cable TV Relay (78) Fixed Microwave (101)
19.7–20.1 FIXED-SATELLITE (space-to-Earth) 5.484A 5.516B Mobile-satellite (space-to-Earth) 5.524	19.7–20.1 FIXED-SATELLITE (space-to-Earth) 5.484A 5.516B MOBILE-SATELLITE (space-to-Earth) 5.524 5.525 5.526 5.527 5.528 5.529	19.7–20.1 FIXED-SATELLITE (space-to-Earth) 5.484A 5.516B Mobile-satellite (space-to-Earth) 5.524		19.7–20.2 FIXED-SATELLITE (space-to-Earth) MOBILE-SATELLITE (space-to-Earth)	Satellite Communications (25)

(Continued)

TABLE F.1 (*Continued*)
Worldwide Frequency Allocations

Table of Frequency Allocations

17.7–23.6 GHz (SHF)

International Table			United States Table		
Region 1 Table	Region 2 Table	Region 3 Table	Federal Table	Non-Federal Table	FCC Rule Part(s)
20.1–20.2					
FIXED-SATELLITE (space-to-Earth) 5.484A 5.516B				5.525 5.526 5.527	
MOBILE-SATELLITE (space-to-Earth)				5.528 5.529	
				US334	
5.524 5.525 5.526 5.527 5.528					
20.2–21.2			20.2–21.2	20.2–21.2	
FIXED-SATELLITE (space-to-Earth)			FIXED-SATELLITE	Standard frequency	
MOBILE-SATELLITE (space-to-Earth)			(space-to-Earth)	and time	
Standard frequency and time signal-satellite (space-to-Earth)			MOBILE-SATELLITE	signal-satellite	
			(space-to-Earth)	(space-to-Earth)	
			Standard frequency		
			and time		
			signal-satellite		
			(space-to-Earth)		
5.524			G117		
21.2–21.4			21.2–21.4		
EARTH EXPLORATION-SATELLITE (passive)			EARTH EXPLORATION-SATELLITE (passive)	EARTH EXPLORATION-SATELLITE (passive)	Fixed Microwave (101)
FIXED			FIXED		
MOBILE			MOBILE		
SPACE RESEARCH (passive)			SPACE RESEARCH (passive)		
			US263		

(Continued)

TABLE F.1 (*Continued*)

Worldwide Frequency Allocations

Table of Frequency Allocations · 17.7–23.6 GHz (SHF)

International Table			United States Table		FCC Rule Part(s)
Region 1 Table	Region 2 Table	Region 3 Table	Federal Table	Non-Federal Table	
21.4–22 FIXED MOBILE BROADCASTING-SATELLITE 5.208B 5.530	21.4–22 FIXED MOBILE	21.4–22 FIXED MOBILE BROADCASTING-SATELLITE 5.208B 5.530 5.531	21.4–22 FIXED MOBILE		
22–22.21 FIXED MOBILE except aeronautical mobile 5.149			22–22.21 FIXED MOBILE except aeronautical mobile US342	MOBILE except aeronautical mobile US342	
22.21–22.5 EARTH EXPLORATION-SATELLITE (passive) FIXED MOBILE except aeronautical mobile RADIO ASTRONOMY SPACE RESEARCH (passive) 5.149 5.532			22.21–22.5 EARTH EXPLORATION-SATELLITE (passive) FIXED MOBILE except aeronautical mobile RADIO ASTRONOMY SPACE RESEARCH (passive) US263 US342	EARTH EXPLORATION-SATELLITE (passive) MOBILE except aeronautical mobile	

(*Continued*)

TABLE F.1 (*Continued*)

Worldwide Frequency Allocations

Table of Frequency Allocations | 17.7–23.6 GHz (SHF)

| | International Table | | United States Table | | |
Region 1 Table	Region 2 Table	Region 3 Table	Federal Table	Non-Federal Table	FCC Rule Part(s)
22.5–22.55 FIXED MOBILE			22.5–22.55 FIXED MOBILE US211		
22.55–23.55 FIXED INTER-SATELLITE 5.338A MOBILE 5.149			22.55–23.55 FIXED INTER-SATELLITE US278 MOBILE US342		Satellite Communications (25) Fixed Microwave (101)
23.55–23.6 FIXED MOBILE			23.55–23.6 FIXED MOBILE		Fixed Microwave (101)
23.6–30 GHz (SHF)					
23.6–24 EARTH EXPLORATION-SATELLITE (passive) RADIO ASTRONOMY SPACE RESEARCH (passive) 5.340			23.6–24 EARTH EXPLORATION-SATELLITE (passive) RADIO ASTRONOMY US74 SPACE RESEARCH (passive) US246		

(*Continued*)

TABLE F.1 (Continued)

Worldwide Frequency Allocations

Table of Frequency Allocations   23.6–30 GHz (SHF)

International Table			United States Table		FCC Rule Part(s)
Region 1 Table	Region 2 Table	Region 3 Table	Federal Table	Non-Federal Table	
24–24.05 AMATEUR AMATEUR-SATELLITE			24-24.05	24-24.05 AMATEUR AMATEUR-SATELLITE	ISM Equipment (18) Amateur Radio (97)
5.150			5.150 US211	5.150 US211	
24.05–24.25 RADIOLOCATION Amateur Earth exploration-satellite (active)			24.05-24.25 RADIOLOCATION G59 Earth exploration-satellite (active)	24.05-24.25 Amateur Earth exploration-satellite (active) Radiolocation	ISM Equipment (18) Private Land Mobile (90) Amateur Radio (97)
5.150			5.150	5.150	
24.25–24.45 FIXED	24.25–24.45 RADIO-NAVIGATION	24.25–24.45 RADIO-NAVIGATION FIXED MOBILE	24.25-24.45	24.25-24.45 FIXED	Fixed Microwave (101)

(Continued)

TABLE F.1 (*Continued*)
Worldwide Frequency Allocations
Table of Frequency Allocations

23.6–30 GHz (SHF)

International Table			United States Table		FCC Rule Part(s)
Region 1 Table	Region 2 Table	Region 3 Table	Federal Table	Non-Federal Table	
24.45–24.75 FIXED INTER-SATELLITE	24.45–24.65 INTER-SATELLITE RADIO- NAVIGATION	24.45–24.65 FIXED INTER-SATELLITE MOBILE RADIO- NAVIGATION	24.45–24.65 INTER-SATELLITE RADIONAVIGATION		Satellite Communications (25)
	5.533	5.533	5.533		
	24.65–24.75 INTER-SATELLITE RADIOLOCATION- SATELLITE (Earth-to-space)	24.65–24.75 FIXED INTER-SATELLITE MOBILE	24.65–24.75 INTER-SATELLITE RADIOLOCATION-SATELLITE (Earth-to-space)		
		5.533	5.533		
24.75–25.25 FIXED	24.75–25.25 FIXED FIXED-SATELLITE (Earth-to-space) 5.535	24.75–25.25 FIXED FIXED-SATELLITE (Earth-to-space) 5.535 MOBILE	24.75–25.05 RADIO- NAVIGATION	24.75–25.05 FIXED-SATELLITE (Earth-to-space) NG167 RADIO- NAVIGATION	Satellite Communications (25) Aviation (87)
			25.05–25.25	25.05–25.25 FIXED FIXED-SATELLITE (Earth-to-space) NG167	Satellite Communications (25) Fixed Microwave (101)

(Continued)

TABLE F.1 (*Continued*)

Worldwide Frequency Allocations

Table of Frequency Allocations   23.6–30 GHz (SHF)

International Table			United States Table		FCC Rule Part(s)
Region 1 Table	Region 2 Table	Region 3 Table	Federal Table	Non-Federal Table	
25.25–25.5 FIXED INTER-SATELLITE 5.536 MOBILE Standard frequency and time signal-satellite (Earth-to-space)			25.25–25.5 FIXED INTER-SATELLITE 5.536 MOBILE Standard frequency and time signal-satellite (Earth-to-space)	25.25–25.5 Inter-satellite 5.536 Standard frequency and time signal-satellite (Earth-to-space)	
25.5–27 EARTH EXPLORATION-SATELLITE (space-to-Earth) 5.536B FIXED INTER-SATELLITE 5.536 MOBILE SPACE RESEARCH (space-to-Earth) 5.536C Standard frequency and time signal-satellite (Earth-to-space) 5.536A			25.5–27 EARTH EXPLORATION-SATELLITE (space-to-Earth) FIXED INTER-SATELLITE 5.536 MOBILE SPACE RESEARCH (space-to-Earth) Standard frequency and time signal-satellite (Earth-to-space) 5.536A US258	25.5–27 Inter-satellite 5.536 Standard frequency and time signal-satellite (Earth-to-space) 5.536A US258	

(*Continued*)

TABLE F.1 (*Continued*)

Worldwide Frequency Allocations

Table of Frequency Allocations — 23.6–30 GHz (SHF)

International Table			United States Table		
Region 1 Table	Region 2 Table	Region 3 Table	Federal Table	Non-Federal Table	FCC Rule Part(s)
27–27.5 FIXED INTER-SATELLITE 5.536 MOBILE	27–27.5 FIXED FIXED-SATELLITE (Earth-to-space) INTER-SATELLITE 5.536 5.537 MOBILE		27–27.5 FIXED INTER-SATELLITE 5.536 MOBILE	27–27.5 Inter-satellite 5.536	
27.5–28.5 FIXED 5.537A FIXED-SATELLITE (Earth-to-space) 5.484A 5.516B 5.539 MOBILE			27.5–30	27.5–29.5 FIXED FIXED-SATELLITE (Earth-to-space) MOBILE	Satellite Communications (25) Fixed Microwave (101)
5.538 5.540					
28.5–29.1 FIXED FIXED-SATELLITE (Earth-to-space) 5.484A 5.516B 5.523A 5.539 MOBILE Earth exploration-satellite (Earth-to-space) 5.541					
5.540					

(Continued)

TABLE F.1 (*Continued*)

Worldwide Frequency Allocations

Table of Frequency Allocations

23.6–30 GHz (SHF)

International Table			United States Table			
Region 1 Table	Region 2 Table	Region 3 Table	Federal Table	Non-Federal Table	FCC Rule Part(s)	
29.1–29.5 FIXED FIXED-SATELLITE (Earth-to-space) 5.516B 5.523C 5.523E 5.535A 5.539 5.541A MOBILE Earth exploration-satellite (Earth-to-space) 5.541						
5.540						
29.5–29.9 FIXED-SATELLITE (Earth-to-space) 5.484A 5.516B 5.539 Earth exploration-satellite (Earth-to-space) 5.541 Mobile-satellite (Earth-to-space)	29.5–29.9 FIXED-SATELLITE (Earth-to-space) 5.484A 5.516B 5.539 MOBILE-SATELLITE (Earth-to-space) Earth exploration-satellite (Earth-to-space) 5.541	29.5–29.9 FIXED-SATELLITE (Earth-to-space) 5.484A 5.516B 5.539 Earth exploration-satellite (Earth-to-space) 5.541 Mobile-satellite (Earth-to-space)		29.5–30 FIXED-SATELLITE (Earth-to-space) MOBILE-SATELLITE (Earth-to-space)	Satellite Communications (25)	
	5.525 5.526 5.527 5.529 5.540					
5.540 5.542	5.542	5.540 5.542				

(*Continued*)

TABLE F.1 (Continued)

Worldwide Frequency Allocations

Table of Frequency Allocations

23.6–30 GHz (SHF)

International Table			United States Table		FCC Rule Part(s)
Region 1 Table	Region 2 Table	Region 3 Table	Federal Table	Non-Federal Table	
29.9–30 FIXED-SATELLITE (Earth-to-space) 5.484A 5.516B 5.539 MOBILE-SATELLITE (Earth-to-space) Earth exploration-satellite (Earth-to-space) 5.541 5.543					
5.525 5.526 5.527 5.538 5.540 5.542				5.525 5.526 5.527 5.529 5.543	
30–39.5 GHz (EHF)					
30–31 FIXED-SATELLITE (Earth-to-space) 5.338A MOBILE-SATELLITE (Earth-to-space) Standard frequency and time signal-satellite (space-to-Earth)			30–31 FIXED-SATELLITE (Earth-to-space) MOBILE-SATELLITE (Earth-to-space) Standard frequency and time signal-satellite (space-to-Earth)	30–31 Standard frequency and time signal-satellite (space-to-Earth)	
5.542			G117		

(Continued)

TABLE F.1 (*Continued*)

Worldwide Frequency Allocations

Table of Frequency Allocations

30–39.5 GHz (EHF)

International Table			United States Table		
Region 1 Table	Region 2 Table	Region 3 Table	Federal Table	Non-Federal Table	FCC Rule Part(s)
31–31.3 FIXED 5.338A 5.543A MOBILE Standard frequency and time signal-satellite (space-to-Earth) Space research 5.544 5.545			31–31.3 Standard frequency and time signal-satellite (space-to-Earth)	31–31.3 FIXED MOBILE Standard frequency and time signal-satellite (space-to-Earth)	Fixed Microwave (101)
5.149			US211 US342	US211 US342	
31.3–31.5 EARTH EXPLORATION-SATELLITE (passive) RADIO ASTRONOMY SPACE RESEARCH (passive)			31.3–31.8 EARTH EXPLORATION-SATELLITE (passive) RADIO ASTRONOMY US74 SPACE RESEARCH (passive)	EARTH EXPLORATION-SATELLITE (passive)	
5.340					

(Continued)

TABLE F.1 (*Continued*)

Worldwide Frequency Allocations

Table of Frequency Allocations

30–39.5 GHz (EHF)

International Table			United States Table		
Region 1 Table	Region 2 Table	Region 3 Table	Federal Table	Non-Federal Table	FCC Rule Part(s)
31.5–31.8 EARTHEXPLORATION-SATELLITE (passive) RADIO ASTRONOMY SPACE RESEARCH (passive) Fixed Mobile except aeronautical mobile	31.5–31.8 EARTH EXPLORATION-SATELLITE (passive) RADIO ASTRONOMY SPACE RESEARCH (passive)	31.5–31.8 EARTH EXPLORATION-SATELLITE (passive) RADIO ASTRONOMY SPACE RESEARCH (passive) Fixed Mobile except aeronautical mobile			
5.149 5.546	5.340	5.149	US246		
31.8–32 FIXED 5.547A RADIONAVIGATION SPACE RESEARCH (deep space) (space-to-Earth)	SPACE RESEARCH (deep space) (space-to-Earth)		31.8–32.3 RADIO-NAVIGATION US69 SPACE RESEARCH (deep space) (space-to-Earth) US262	31.8–32.3 SPACE RESEARCH (deep space) (space-to-Earth) US262	
5.547 5.547B 5.548					

(*Continued*)

TABLE F.1 (Continued)
Worldwide Frequency Allocations

Table of Frequency Allocations

30–39.5 GHz (EHF)

International Table			United States Table		FCC Rule Part(s)
Region 1 Table	Region 2 Table	Region 3 Table	Federal Table	Non-Federal Table	
32–32.3 FIXED 5.547A RADIONAVIGATION SPACE RESEARCH (deep space) (space-to-Earth)					
5.547 5.547C 5.548			5.548 US211	5.548 US211	
32.3–33 FIXED 5.547A INTER-SATELLITE RADIONAVIGATION			32.3–33 INTER-SATELLITE US278 RADIONAVIGATION US69		Aviation (87)
5.547 5.547D 5.548			5.548		
33–33.4 FIXED 5.547A RADIONAVIGATION			33–33.4 RADIONAVIGATION US69		
5.547 5.547E			US360 G117		
33.4–34.2 RADIOLOCATION			33.4–34.2 RADIOLOCATION	33.4–34.2 Radiolocation	Private Land Mobile (90)
5.549			US360 G117	US360	

(Continued)

TABLE F.1 (*Continued*)

Worldwide Frequency Allocations

Table of Frequency Allocations

30–39.5 GHz (EHF)

International Table			United States Table		FCC Rule Part(s)
Region 1 Table	Region 2 Table	Region 3 Table	Federal Table	Non-Federal Table	
34.2–34.7 RADIOLOCATION SPACE RESEARCH (deep space) (Earth-to-space)	RADIOLOCATION SPACE RESEARCH (deep space) (Earth-to-space)		34.2–34.7 RADIOLOCATION SPACE RESEARCH (deep space) (Earth-to-space) US262 US360 G34 G117	34.2–34.7 Radiolocation Space research (deep space) (Earth-to-space) US262 US360	
5.549			34.7–35.5 RADIOLOCATION	34.7–35.5 Radiolocation	
34.7–35.2 RADIOLOCATION Space research 5.550					
5.549					
35.2–35.5 METEOROLOGICAL AIDS RADIOLOCATION			US360 G117	US360	
5.549					

(Continued)

TABLE F.1 (*Continued*)

Worldwide Frequency Allocations

Table of Frequency Allocations

30–39.5 GHz (EHF)

International Table			United States Table		FCC Rule Part(s)
Region 1 Table	Region 2 Table	Region 3 Table	Federal Table	Non-Federal Table	
35.5–36 METEOROLOGICAL AIDS EARTH EXPLORATION-SATELLITE (active) RADIOLOCATION SPACE RESEARCH (active)			35.5–36 EARTH EXPLORATION-SATELLITE (active) RADIOLOCATION SPACE RESEARCH (active)	35.5–36 Earth exploration-satellite (active) Radiolocation Space research (active)	
5.549 5.549A			US360 G117	US360	
36–37 EARTH EXPLORATION-SATELLITE (passive) FIXED MOBILE SPACE RESEARCH (passive)			36–37 EARTH EXPLORATION-SATELLITE (passive) FIXED MOBILE SPACE RESEARCH (passive)		
5.149 5.550A			US263 US342		
37–37.5 FIXED MOBILE SPACE RESEARCH (space-to-Earth)			37–38 FIXED MOBILE SPACE RESEARCH (space-to-Earth)	37–37.5 FIXED MOBILE	
5.547					

(Continued)

TABLE F.1 (Continued)
Worldwide Frequency Allocations

Table of Frequency Allocations

30–39.5 GHz (EHF)

International Table			United States Table		FCC Rule Part(s)
Region 1 Table	Region 2 Table	Region 3 Table	Federal Table	Non-Federal Table	
37.5–38 FIXED FIXED-SATELLITE (space-to-Earth) MOBILE SPACE RESEARCH (space-to-Earth) Earth exploration-satellite (space-to-Earth)				37.5–38.6 FIXED FIXED-SATELLITE (space-to-Earth) MOBILE	Satellite Communications (25)
5.547 38–39.5 FIXED FIXED-SATELLITE (space-to-Earth) MOBILE Earth exploration-satellite (space-to-Earth)			38–38.6 FIXED MOBILE 38.6–39.5	38.6–39.5 FIXED FIXED-SATELLITE (space-to-Earth) MOBILE NG175	Satellite Communications (25) Fixed Microwave (101)

(Continued)

TABLE F.1 (*Continued*)

Worldwide Frequency Allocations

Table of Frequency Allocations

| | | | 39.5–50.2 GHz (EHF) | | |
| International Table | | | United States Table | | |
Region 1 Table	Region 2 Table	Region 3 Table	Federal Table	Non-Federal Table	FCC Rule Part(s)
39.5–40 FIXED FIXED-SATELLITE (space-to-Earth) 5.516B MOBILE MOBILE-SATELLITE (space-to-Earth) Earth exploration-satellite (space-to-Earth)			39.5–40 FIXED-SATELLITE (space-to-Earth) MOBILE-SATELLITE (space-to-Earth) US382	39.5–40 FIXED FIXED-SATELLITE (space-to-Earth) MOBILE NG175	Satellite Communications (25) Fixed Microwave (101)
5.547 40–40.5 EARTH EXPLORATION-SATELLITE (Earth-to-space) FIXED FIXED-SATELLITE (space-to-Earth) 5.516B MOBILE MOBILE-SATELLITE (space-to-Earth) SPACE RESEARCH (Earth-to-space) Earth exploration-satellite (space-to-Earth)			G117 40–40.5 EARTH EXPLORATION-SATELLITE (Earth-to-space) FIXED-SATELLITE (space-to-Earth) MOBILE-SATELLITE (space-to-Earth) SPACE RESEARCH (Earth-to-space) Earth exploration-satellite (space-to-Earth) G117	US382 40–40.5 FIXED-SATELLITE (space-to-Earth) MOBILE-SATELLITE (space-to-Earth)	Satellite Communications (25)

(Continued)

TABLE F.1 (*Continued*)

Worldwide Frequency Allocations

Table of Frequency Allocations

39.5–50.2 GHz (EHF)

International Table			United States Table		FCC Rule Part(s)
Region 1 Table	Region 2 Table	Region 3 Table	Federal Table	Non-Federal Table	
40.5–41 FIXED FIXED-SATELLITE (space-to-Earth) BROADCASTING BROADCASTING-SATELLITE Mobile 5.547	40.5–41 FIXED FIXED-SATELLITE (space-to-Earth) 5.516B BROADCASTING BROADCASTING-SATELLITE Mobile Mobile-satellite (space-to-Earth) 5.547	40.5–41 FIXED FIXED-SATELLITE (space-to-Earth) BROADCASTING BROADCASTING-SATELLITE Mobile 5.547	40.5–41 FIXED-SATELLITE (space-to-Earth) Mobile-satellite (space-to-Earth) US211 G117	40.5–41 FIXED-SATELLITE (space-to-Earth) BROADCASTING BROADCASTING-SATELLITE Fixed Mobile Mobile-satellite (space-to-Earth) US211	
41–42.5 FIXED FIXED-SATELLITE (space-to-Earth) 5.516B BROADCASTING BROADCASTING-SATELLITE Mobile			41–42.5	41–42 FIXED FIXED-SATELLITE (space-to-Earth) MOBILE BROADCASTING BROADCASTING-SATELLITE US211	

(*Continued*)

TABLE F.1 (Continued)

Worldwide Frequency Allocations

Table of Frequency Allocations   39.5–50.2 GHz (EHF)

International Table			United States Table		FCC Rule Part(s)
Region 1 Table	Region 2 Table	Region 3 Table	Federal Table	Non-Federal Table	
5.547 5.551F 5.551H 5.551I			US211	42–42.5 FIXED MOBILE BROADCASTING BROADCASTING-SATELLITE US211	
42.5–43.5 FIXED FIXED-SATELLITE (Earth-to-space) 5.552 MOBILE except aeronautical mobile RADIO ASTRONOMY			42.5–43.5 FIXED FIXED-SATELLITE (Earth-to-space) MOBILE except aeronautical mobile RADIO ASTRONOMY	42.5–43.5 RADIO ASTRONOMY	
5.149 5.547			US342	US342	
43.5–47 MOBILE 5.553 MOBILE-SATELLITE RADIONAVIGATION RADIONAVIGATION-SATELLITE			43.5–45.5 FIXED-SATELLITE (Earth-to-space) MOBILE-SATELLITE (Earth-to-space) G117	43.5–45.5	

(Continued)

TABLE F.1 (*Continued*)

Worldwide Frequency Allocations

Table of Frequency Allocations 39.5–50.2 GHz (EHF)

International Table			United States Table		FCC Rule Part(s)
Region 1 Table	Region 2 Table	Region 3 Table	Federal Table	Non-Federal Table	
			45.5–46.9 MOBILE MOBILE-SATELLITE (Earth-to-space) RADIONAVIGATION-SATELLITE 5.554		RF Devices (15)
			46.9–47 MOBILE MOBILE-SATELLITE (Earth-to-space) RADIO-NAVIGATION-SATELLITE 5.554	46.9–47 FIXED MOBILE MOBILE-SATELLITE (Earth-to-space) RADIO-NAVIGATION-SATELLITE 5.554	
5.554 47–47.2 AMATEUR AMATEUR-SATELLITE			47–48.2	47–47.2 AMATEUR AMATEUR-SATELLITE	Amateur Radio (97)
47.2–47.5 FIXED FIXED-SATELLITE (Earth-to-space) 5.552 MOBILE 5.552A	47.2–47.5 FIXED FIXED-SATELLITE (Earth-to-space) 5.552			47.2–48.2 FIXED FIXED-SATELLITE (Earth-to-space) US297 MOBILE	Satellite Communications (25)

(*Continued*)

TABLE F.1 (*Continued*)

Worldwide Frequency Allocations

Table of Frequency Allocations   39.5–50.2 GHz (EHF)

International Table			United States Table		FCC Rule Part(s)
Region 1 Table	Region 2 Table	Region 3 Table	Federal Table	Non-Federal Table	
47.5–47.9 FIXED FIXED-SATELLITE (Earth-to-space) 5.552 (space-to-Earth) 5.516B 5.554A MOBILE	47.5–47.9 FIXED FIXED-SATELLITE (Earth-to-space) 5.552 MOBILE	FIXED-SATELLITE (Earth-to-space) 5.552 MOBILE			
47.9–48.2 FIXED FIXED-SATELLITE (Earth-to-space) 5.552 MOBILE		FIXED-SATELLITE (Earth-to-space) 5.552			
5.552A					
48.2–48.54 FIXED FIXED-SATELLITE (Earth-to-space) 5.552 (space-to-Earth) 5.516B 5.554A 5.555B MOBILE	48.2–50.2 FIXED FIXED-SATELLITE (Earth-to-space) 5.338A 5.516B 5.552 MOBILE		48.2–50.2 FIXED FIXED-SATELLITE (Earth-to-space) US264 MOBILE US264	FIXED-SATELLITE (Earth-to-space) US297	

(*Continued*)

TABLE F.1 (*Continued*)

Worldwide Frequency Allocations

Table of Frequency Allocations

50.2–71 GHz (EHF)

International Table			United States Table		FCC Rule Part(s)
Region 1 Table	Region 2 Table	Region 3 Table	Federal Table	Non-Federal Table	
48.54–49.44 FIXED FIXED-SATELLITE (Earth-to-space) 5.552 MOBILE	5.149 5.340 5.555		5.555 US342		
49.44–50.2 FIXED FIXED-SATELLITE (Earth-to-space) 5.338A 5.552 (space-to-Earth) 5.516B 5.554A 5.555B MOBILE					
50.2–50.4 EARTH EXPLORATION-SATELLITE (passive) SPACE RESEARCH (passive) 5.340			50.2–50.4 EARTH EXPLORATION-SATELLITE (passive) SPACE RESEARCH (passive) US246		
50.4–51.4 FIXED FIXED-SATELLITE (Earth-to-space) 5.338A MOBILE Mobile-satellite (Earth-to-space)			50.4–51.4 FIXED FIXED-SATELLITE (Earth-to-space) MOBILE MOBILE-SATELLITE (Earth-to-space) G117	50.4–51.4 FIXED FIXED-SATELLITE (Earth-to-space) MOBILE MOBILE-SATELLITE (Earth-to-space)	

(Continued)

TABLE F.1 (Continued)

Worldwide Frequency Allocations

Table of Frequency Allocations   50.2–71 GHz (EHF)

International Table			United States Table		FCC Rule Part(s)
Region 1 Table	Region 2 Table	Region 3 Table	Federal Table	Non-Federal Table	
51.4–52.6 FIXED 5.338A MOBILE 5.547 5.556			51.4–52.6 FIXED MOBILE		
52.6–54.25 EARTH EXPLORATION-SATELLITE (passive) SPACE RESEARCH (passive) 5.340 5.556			52.6–54.25 EARTH EXPLORATION-SATELLITE (passive) SPACE RESEARCH (passive) US246		
54.25–55.78 EARTH EXPLORATION-SATELLITE (passive) INTER-SATELLITE 5.556A SPACE RESEARCH (passive) 5.556B			54.25–55.78 EARTH EXPLORATION-SATELLITE (passive) INTER-SATELLITE 5.556A SPACE RESEARCH (passive)		Satellite Communications (25)
55.78–56.9 EARTH EXPLORATION-SATELLITE (passive) FIXED 5.557A INTER-SATELLITE 5.556A MOBILE 5.558 SPACE RESEARCH (passive) 5.547 5.557			55.78–56.9 EARTH EXPLORATION-SATELLITE (passive) FIXED US379 INTER-SATELLITE 5.558 MOBILE 5.558 SPACE RESEARCH (passive) US263 US353		

(Continued)

TABLE F.1 (*Continued*)
Worldwide Frequency Allocations

Table of Frequency Allocations | 50.2–71 GHz (EHF)

International Table			United States Table			FCC Rule Part(s)
Region 1 Table	Region 2 Table	Region 3 Table	Federal Table	Non-Federal Table		
56.9–57	56.9–57		56.9–57	56.9–57		
EARTH EXPLORATION-SATELLITE (passive)			EARTH EXPLORATION-SATELLITE (passive)	EARTH EXPLORATION-SATELLITE (passive)		
FIXED			FIXED	FIXED		
INTER-SATELLITE 5.558A			INTER-SATELLITE	MOBILE 5.558		
MOBILE 5.558			G128	SPACE RESEARCH (passive)		
SPACE RESEARCH (passive)			MOBILE 5.558			
			SPACE RESEARCH (passive)			
			US263	US263		
5.547 5.557						
57–58.2			57–58.2			RF Devices (15)
EARTH EXPLORATION-SATELLITE (passive)			EARTH EXPLORATION-SATELLITE (passive)			Satellite
FIXED			FIXED			Communications (25)
INTER-SATELLITE 5.556A			INTER-SATELLITE 5.556A			
MOBILE 5.558			MOBILE 5.558			
SPACE RESEARCH (passive)			SPACE RESEARCH (passive)			
5.547 5.557			US263			

(Continued)

TABLE F.1 (*Continued*)

Worldwide Frequency Allocations

Table of Frequency Allocations

50.2–71 GHz (EHF)

International Table			United States Table		FCC Rule Part(s)
Region 1 Table	Region 2 Table	Region 3 Table	Federal Table	Non-Federal Table	
58.2–59 EARTH EXPLORATION-SATELLITE (passive) FIXED MOBILE SPACE RESEARCH (passive) 5.547 5.556			58.2–59 EARTH EXPLORATION-SATELLITE (passive) FIXED MOBILE SPACE RESEARCH (passive) US353 US354		RF Devices (15)
59–59.3 EARTH EXPLORATION-SATELLITE (passive) FIXED INTER-SATELLITE 5.556A MOBILE 5.558 RADIOLOCATION 5.559 SPACE RESEARCH (passive)			59–59.3 EARTH EXPLORATION- SATELLITE (passive) FIXED INTER-SATELLITE 5.556A MOBILE 5.558 RADIOLOCATION 5.559 SPACE RESEARCH (passive) US353	59–59.3 EARTH EXPLORATION- SATELLITE (passive) FIXED MOBILE 5.558 RADIOLOCATION 5.559 SPACE RESEARCH (passive) US353	

(Continued)

TABLE F.1 (*Continued*)
Worldwide Frequency Allocations
Table of Frequency Allocations

	International Table			United States Table		
				50.2–71 GHz (EHF)		
Region 1 Table	Region 2 Table	Region 3 Table	Federal Table	Non-Federal Table	FCC Rule Part(s)
59.3–64 FIXED INTER-SATELLITE MOBILE 5.558 RADIOLOCATION 5.559			59.3–64 FIXED INTER-SATELLITE MOBILE 5.558 RADIOLOCATION 5.559	59.3–64 FIXED MOBILE 5.558 RADIOLOCATION 5.559	RF Devices (15) ISM Equipment (18)
5.138			5.138 US353	5.138 US353	
64–65 FIXED INTER-SATELLITE MOBILE except aeronautical mobile 5.547 5.556			64–65 FIXED INTER-SATELLITE MOBILE except aeronautical mobile	64–65 FIXED MOBILE except aeronautical mobile	
65–66 EARTH EXPLORATION-SATELLITE FIXED INTER-SATELLITE MOBILE except aeronautical mobile SPACE RESEARCH			65–66 EARTH EXPLORATION-SATELLITE FIXED MOBILE except aeronautical mobile SPACE RESEARCH	65–66 EARTH EXPLORATION-SATELLITE FIXED INTER-SATELLITE MOBILE except aeronautical mobile SPACE RESEARCH	Satellite Communications (25)
5.547					

(*Continued*)

TABLE F.1 (Continued)

Worldwide Frequency Allocations

Table of Frequency Allocations

50.2–71 GHz (EHF)

International Table			United States Table		FCC Rule Part(s)
Region 1 Table	Region 2 Table	Region 3 Table	Federal Table	Non-Federal Table	
66–71 INTER-SATELLITE MOBILE 5.553 5.558 MOBILE-SATELLITE RADIONAVIGATION RADIONAVIGATION-SATELLITE 5.554			66–71 MOBILE 5.553 5.558 MOBILE-SATELLITE RADIO-NAVIGATION RADIO-NAVIGATION-SATELLITE 5.554	66–71 INTER-SATELLITE MOBILE 5.553 5.558 MOBILE-SATELLITE RADIO-NAVIGATION RADIO-NAVIGATION-SATELLITE 5.554	

Table of Frequency Allocations

71–100 GHz (EHF)

Region 1 Table	Region 2 Table	Region 3 Table	Federal Table	Non-Federal Table	FCC Rule Part(s)
71–74 FIXED FIXED-SATELLITE (space-to-Earth) MOBILE MOBILE-SATELLITE (space-to-Earth)			71–74 FIXED FIXED-SATELLITE (space-to-Earth) MOBILE MOBILE-SATELLITE (space-to-Earth) US389		Fixed Microwave (101)

(Continued)

TABLE F.1 (*Continued*)

Worldwide Frequency Allocations

Table of Frequency Allocations

71–100 GHz (EHF)

International Table			United States Table		FCC Rule Part(s)
Region 1 Table	Region 2 Table	Region 3 Table	Federal Table	Non-Federal Table	
74–76 FIXED FIXED-SATELLITE (space-to-Earth) MOBILE BROADCASTING BROADCASTING-SATELLITE Space research (space-to-Earth)			74–76 FIXED FIXED-SATELLITE (space-to-Earth) MOBILE Space research (space-to-Earth) US389	74–76 FIXED FIXED-SATELLITE (space-to-Earth) MOBILE BROADCASTING BROADCASTING- SATELLITE Space research (space-to-Earth) US389	
5.561					
76–77.5 RADIO ASTRONOMY RADIOLOCATION Amateur Amateur-satellite Space research (space-to-Earth)			76–77.5 RADIO ASTRONOMY RADIOLOCATION Space research (space-to-Earth)	76–77 RADIO ASTRONOMY RADIOLOCATION Amateur Space research (space-to-Earth) US342	RF Devices (15)

(*Continued*)

TABLE F.1 (*Continued*)

Worldwide Frequency Allocations

Table of Frequency Allocations

71–100 GHz (EHF)

International Table			United States Table		FCC Rule Part(s)
Region 1 Table	Region 2 Table	Region 3 Table	Federal Table	Non-Federal Table	
5.149			US342	77–77.5 RADIO ASTRONOMY RADIOLOCATION Amateur Amateur-satellite Space research (space-to-Earth) US342	Amateur Radio (97)
77.5–78 AMATEUR AMATEUR-SATELLITE Radio astronomy Space research (space-to-Earth)			77.5–78 Radio astronomy Space research (space-to-Earth)	77.5–78 AMATEUR AMATEUR-SATELLITE Radio astronomy Space research (space-to-Earth)	
5.149			US342	US342	

(*Continued*)

TABLE F.1 (*Continued*)
Worldwide Frequency Allocations

Table of Frequency Allocations

71–100 GHz (EHF)

International Table			United States Table		FCC Rule Part(s)
Region 1 Table	Region 2 Table	Region 3 Table	Federal Table	Non-Federal Table	
78–79 RADIOLOCATION Amateur Amateur-satellite Radio astronomy Space research (space-to-Earth)			78–79 RADIO ASTRONOMY RADIOLOCATION Space research (space-to-Earth)	78–79 RADIO ASTRONOMY RADIOLOCATION Amateur Amateur-satellite Space research (space-to-Earth)	
5.149 5.560			5.560 US342	5.560 US342	
79–81 RADIO ASTRONOMY RADIOLOCATION Amateur Amateur-satellite Space research (space-to-Earth)			79–81 RADIO ASTRONOMY RADIOLOCATION Space research (space-to-Earth)	79–81 RADIO ASTRONOMY RADIOLOCATION Amateur Amateur-satellite Space research (space-to-Earth)	
5.149			US342	US342	

(*Continued*)

TABLE F.1 (*Continued*)

Worldwide Frequency Allocations

Table of Frequency Allocations

71–100 GHz (EHF)

International Table			United States Table		FCC Rule Part(s)
Region 1 Table	Region 2 Table	Region 3 Table	Federal Table	Non-Federal Table	
81–84 FIXED FIXED-SATELLITE (Earth-to-space) MOBILE MOBILE-SATELLITE (Earth-to-space) RADIO ASTRONOMY Space research (space-to-Earth) 5.149 5.561A			81–84 FIXED FIXED-SATELLITE (Earth-to-space) US297 MOBILE MOBILE-SATELLITE (Earth-to-space) RADIO ASTRONOMY Space research (space-to-Earth) US342 US388 US389		Fixed Microwave (101)
84–86 FIXED FIXED-SATELLITE (Earth-to-space) 5.561B MOBILE RADIO ASTRONOMY 5.149			84–86 FIXED FIXED-SATELLITE (Earth-to-space) MOBILE RADIO ASTRONOMY US342 US388 US389		
86–92 EARTH EXPLORATION-SATELLITE (passive) RADIO ASTRONOMY SPACE RESEARCH (passive) 5.340			86–92 EARTH EXPLORATION-SATELLITE US74 RADIO ASTRONOMY SPACE RESEARCH (passive) US246	EARTH EXPLORATION-SATELLITE (passive)	

(*Continued*)

TABLE F.1 (*Continued*)
Worldwide Frequency Allocations

Table of Frequency Allocations

71–100 GHz (EHF)

International Table			United States Table		FCC Rule Part(s)
Region 1 Table	Region 2 Table	Region 3 Table	Federal Table	Non-Federal Table	
92–94 FIXED MOBILE RADIO ASTRONOMY RADIOLOCATION			92–94 FIXED MOBILE RADIO ASTRONOMY RADIOLOCATION		RF Devices (15) Fixed Microwave (101)
5.149			US342 US388		
94–94.1 EARTH EXPLORATION-SATELLITE (active) RADIOLOCATION SPACE RESEARCH (active) Radio astronomy			94–94.1 EARTH EXPLORATION- SATELLITE (active) RADIOLOCATION SPACE RESEARCH (active) Radio astronomy 5.562 5.562A	94–94.1 RADIOLOCATION Radio astronomy	RF Devices (15)
5.562 5.562A				5.562A	
94.1–95 FIXED MOBILE RADIO ASTRONOMY RADIOLOCATION			94.1–95 FIXED MOBILE RADIO ASTRONOMY RADIOLOCATION		RF Devices (15) Fixed Microwave (101)
5.149			US342 US388		

(*Continued*)

TABLE F.1 (*Continued*)

Worldwide Frequency Allocations

Table of Frequency Allocations   71–100 GHz (EHF)

International Table			United States Table		FCC Rule Part(s)
Region 1 Table	Region 2 Table	Region 3 Table	Federal Table	Non-Federal Table	
95–100 FIXED MOBILE RADIO ASTRONOMY RADIOLOCATION RADIONAVIGATION RADIONAVIGATION-SATELLITE 5.340 5.554			95–100 FIXED MOBILE RADIO ASTRONOMY RADIOLOCATION RADIONAVIGATION RADIONAVIGATION-SATELLITE 5.554 US342		

Table of Frequency Allocations   100–155.5 GHz (EHF)

International Table			United States Table		FCC Rule Part(s)
100–102 EARTH EXPLORATION-SATELLITE (passive) RADIO ASTRONOMY SPACE RESEARCH (passive) 5.340 5.341			100–102 EARTH EXPLORATION-SATELLITE (passive) RADIO ASTRONOMY US74 SPACE RESEARCH (passive) 5.341 US246		
102–105 FIXED MOBILE RADIO ASTRONOMY 5.149 5.341			102–105 FIXED MOBILE RADIO ASTRONOMY 5.341 US342		

(*Continued*)

TABLE F.1 (Continued)

Worldwide Frequency Allocations

Table of Frequency Allocations   100–155.5 GHz (EHF)

International Table			United States Table		FCC Rule Part(s)
Region 1 Table	Region 2 Table	Region 3 Table	Federal Table	Non-Federal Table	
105–109.5 FIXED MOBILE RADIO ASTRONOMY SPACE RESEARCH (passive) 5.562B 5.149 5.341			105–109.5 FIXED MOBILE RADIO ASTRONOMY SPACE RESEARCH (passive) 5.562B 5.341 US342		
109.5–111.8 EARTH EXPLORATION-SATELLITE (passive) RADIO ASTRONOMY SPACE RESEARCH (passive) 5.340 5.341			109.5–111.8 EARTH EXPLORATION-SATELLITE (passive) RADIO ASTRONOMY US74 SPACE RESEARCH (passive) 5.341 US246		
111.8–114.25 FIXED MOBILE RADIO ASTRONOMY SPACE RESEARCH (passive) 5.562B 5.149 5.341			111.8–114.25 FIXED MOBILE RADIO ASTRONOMY SPACE RESEARCH (passive) 5.562B 5.341 US342		
114.25–116 EARTH EXPLORATION-SATELLITE (passive) RADIO ASTRONOMY SPACE RESEARCH (passive) 5.340 5.341			114.25–116 EARTH EXPLORATION-SATELLITE (passive) RADIO ASTRONOMY US74 SPACE RESEARCH (passive) 5.341 US246		

(Continued)

TABLE F.1 (*Continued*)

Worldwide Frequency Allocations

Table of Frequency Allocations   100–155.5 GHz (EHF)

International Table			United States Table		FCC Rule Part(s)
Region 1 Table	Region 2 Table	Region 3 Table	Federal Table	Non-Federal Table	
116–119.98 EARTH EXPLORATION-SATELLITE (passive) INTER-SATELLITE 5.562C SPACE RESEARCH (passive) 5.341			116–122.25 EARTH EXPLORATION-SATELLITE (passive) INTER-SATELLITE 5.562C SPACE RESEARCH (passive)	116–122.25 EARTH EXPLORATION-SATELLITE (passive) INTER-SATELLITE 5.562C SPACE RESEARCH (passive)	ISM Equipment (18)
119.98–122.25 EARTH EXPLORATION-SATELLITE (passive) INTER-SATELLITE 5.562C SPACE RESEARCH (passive) 5.138 5.341			5.138 5.341 US211		
122.25–123 FIXED INTER-SATELLITE MOBILE 5.558 Amateur 5.138			122.25–123 FIXED INTER-SATELLITE MOBILE 5.558 5.138	122.25–123 FIXED INTER-SATELLITE MOBILE 5.558 Amateur 5.138	ISM Equipment (18) Amateur Radio (97)
123–130 FIXED-SATELLITE (space-to-Earth) MOBILE-SATELLITE (space-to-Earth) RADIONAVIGATION RADIONAVIGATION-SATELLITE Radio astronomy 5.562D 5.149 5.554			123–130 FIXED-SATELLITE (space-to-Earth) MOBILE-SATELLITE (space-to-Earth) RADIONAVIGATION RADIONAVIGATION-SATELLITE Radio astronomy 5.554 US211 US342		

(Continued)

TABLE F.1 (*Continued*)

Worldwide Frequency Allocations

Table of Frequency Allocations

100–155.5 GHz (EHF)

International Table			United States Table		FCC Rule Part(s)
Region 1 Table	Region 2 Table	Region 3 Table	Federal Table	Non-Federal Table	
130–134 EARTH EXPLORATION-SATELLITE (active) 5.562E FIXED INTER-SATELLITE MOBILE 5.558 RADIO ASTRONOMY 5.149 5.562A			130–134 EARTH EXPLORATION-SATELLITE (active) 5.562E FIXED INTER-SATELLITE MOBILE 5.558 RADIO ASTRONOMY 5.562A US342		
134–136 AMATEUR AMATEUR-SATELLITE Radio astronomy			134–136 Radio astronomy	134–136 AMATEUR AMATEUR-SATELLITE Radio astronomy	Amateur Radio (97)
136–141 RADIO ASTRONOMY RADIOLOCATION Amateur Amateur-satellite 5.149			136–141 RADIO ASTRONOMY RADIOLOCATION US342	136–141 RADIO ASTRONOMY RADIOLOCATION Amateur Amateur-satellite US342	

(*Continued*)

TABLE F.1 (Continued)

Worldwide Frequency Allocations

Table of Frequency Allocations   100–155.5 GHz (EHF)

International Table			United States Table		
Region 1 Table	Region 2 Table	Region 3 Table	Federal Table	Non-Federal Table	FCC Rule Part(s)
141–148.5 FIXED MOBILE RADIO ASTRONOMY RADIOLOCATION 5.149			141–148.5 FIXED MOBILE RADIO ASTRONOMY RADIOLOCATION US342		
148.5–151.5 EARTH EXPLORATION-SATELLITE (passive) RADIO ASTRONOMY SPACE RESEARCH (passive) 5.340			148.5–151.5 EARTH EXPLORATION-SATELLITE (passive) RADIO ASTRONOMY US74 SPACE RESEARCH (passive) US246		
151.5–155.5 FIXED MOBILE RADIO ASTRONOMY RADIOLOCATION 5.149			151.5–155.5 FIXED MOBILE RADIO ASTRONOMY RADIOLOCATION US342		

Table of Frequency Allocations   155.5–238 GHz (EHF)

155.5–158.5 EARTH EXPLORATION-SATELLITE (passive) FIXED MOBILE RADIO ASTRONOMY SPACE RESEARCH (passive) 5.562B 5.149 5.562F 5.562G			155.5–158.5 EARTH EXPLORATION-SATELLITE (passive) FIXED MOBILE RADIO ASTRONOMY SPACE RESEARCH (passive) 5.562G US342		

(Continued)

TABLE F.1 (*Continued*)

Worldwide Frequency Allocations

Table of Frequency Allocations

155.5–238 GHz (EHF)

International Table			United States Table		FCC Rule Part(s)
Region 1 Table	Region 2 Table	Region 3 Table	Federal Table	Non-Federal Table	
158.5–164 FIXED FIXED-SATELLITE (space-to-Earth) MOBILE MOBILE-SATELLITE (space-to-Earth)			158.5–164 FIXED FIXED-SATELLITE (space-to-Earth) MOBILE MOBILE-SATELLITE (space-to-Earth) US211		
164–167 EARTH EXPLORATION-SATELLITE (passive) RADIO ASTRONOMY SPACE RESEARCH (passive) 5.340			164–167 EARTH EXPLORATION-SATELLITE (passive) RADIO ASTRONOMY US74 SPACE RESEARCH (passive) US246		
167–174.5 FIXED FIXED-SATELLITE (space-to-Earth) INTER-SATELLITE MOBILE 5.558 5.149 5.562D			167–174.5 FIXED FIXED-SATELLITE (space-to-Earth) INTER-SATELLITE MOBILE 5.558 US211 US342		
174.5–174.8 FIXED INTER-SATELLITE MOBILE 5.558			174.5–174.8 FIXED INTER-SATELLITE MOBILE 5.558		

(*Continued*)

TABLE F.1 (*Continued*)

Worldwide Frequency Allocations

Table of Frequency Allocations

155.5–238 GHz (EHF)

International Table			United States Table		FCC Rule Part(s)
Region 1 Table	Region 2 Table	Region 3 Table	Federal Table	Non-Federal Table	
174.8–182 EARTH EXPLORATION-SATELLITE (passive) INTER-SATELLITE 5.562H SPACE RESEARCH (passive)			174.8–182 EARTH EXPLORATION-SATELLITE (passive) INTER-SATELLITE 5.562H SPACE RESEARCH (passive)	EARTH EXPLORATION-SATELLITE (passive)	
182–185 EARTH EXPLORATION-SATELLITE (passive) RADIO ASTRONOMY SPACE RESEARCH (passive) 5.340			182–185 EARTH EXPLORATION-SATELLITE (passive) RADIO ASTRONOMY SPACE RESEARCH (passive) US246	EARTH EXPLORATION-SATELLITE (passive)	
185–190 EARTH EXPLORATION-SATELLITE (passive) INTER-SATELLITE 5.562H SPACE RESEARCH (passive)			185–190 EARTH EXPLORATION-SATELLITE (passive) INTER-SATELLITE 5.562H SPACE RESEARCH (passive)	EARTH EXPLORATION-SATELLITE (passive)	
190–191.8 EARTH EXPLORATION-SATELLITE (passive) SPACE RESEARCH (passive) 5.340			190–191.8 EARTH EXPLORATION-SATELLITE (passive) SPACE RESEARCH (passive) US246	EARTH EXPLORATION-SATELLITE (passive)	
191.8–200 FIXED INTER-SATELLITE MOBILE 5.558 MOBILE-SATELLITE RADIONAVIGATION RADIONAVIGATION-SATELLITE 5.149 5.341 5.554			191.8–200 FIXED INTER-SATELLITE MOBILE 5.558 MOBILE-SATELLITE RADIONAVIGATION RADIONAVIGATION-SATELLITE 5.341 5.554 US211 US342		

(*Continued*)

TABLE F.1 (*Continued*)

Worldwide Frequency Allocations

Table of Frequency Allocations

155.5–238 GHz (EHF)

International Table			United States Table		FCC Rule Part(s)
Region 1 Table	Region 2 Table	Region 3 Table	Federal Table	Non-Federal Table	
200–209 EARTH EXPLORATION-SATELLITE (passive) RADIO ASTRONOMY SPACE RESEARCH (passive) 5.340 5.341 5.563A			200–209 EARTH EXPLORATION-SATELLITE (passive) RADIO ASTRONOMY US74 SPACE RESEARCH (passive) 5.341 5.563A US246	200–209 EARTH EXPLORATION-SATELLITE (passive) RADIO ASTRONOMY US74 SPACE RESEARCH (passive) 5.341 5.563A US246	
209–217 FIXED FIXED-SATELLITE (Earth-to-space) MOBILE RADIO ASTRONOMY 5.149 5.341			209–217 FIXED FIXED-SATELLITE (Earth-to-space) MOBILE RADIO ASTRONOMY 5.341 US342		
217–226 FIXED FIXED-SATELLITE (Earth-to-space) MOBILE RADIO ASTRONOMY SPACE RESEARCH (passive) 5.562B 5.149 5.341			217–226 FIXED FIXED-SATELLITE (Earth-to-space) MOBILE RADIO ASTRONOMY SPACE RESEARCH (passive) 5.562B 5.341 US342		
226–231.5 EARTH EXPLORATION-SATELLITE (passive) RADIO ASTRONOMY SPACE RESEARCH (passive) 5.340			226–231.5 EARTH EXPLORATION-SATELLITE (passive) RADIO ASTRONOMY SPACE RESEARCH (passive) US246		

(*Continued*)

TABLE F.1 (*Continued*)

Worldwide Frequency Allocations

Table of Frequency Allocations

International Table			United States Table		FCC Rule Part(s)
Region 1 Table	Region 2 Table	Region 3 Table	Federal Table	Non-Federal Table	
			155.5–238 GHz (EHF)		
231.5–232 FIXED MOBILE Radiolocation			231.5–232 FIXED MOBILE Radiolocation		
232–235 FIXED FIXED-SATELLITE (space-to-Earth) MOBILE Radiolocation			232–235 FIXED FIXED-SATELLITE (space-to-Earth) MOBILE Radiolocation		
235–238 EARTH EXPLORATION-SATELLITE (passive) FIXED-SATELLITE (space-to-Earth) SPACE RESEARCH (passive) 5.563A 5.563B			235–238 EARTH EXPLORATION-SATELLITE (passive) FIXED-SATELLITE (space-to-Earth) SPACE RESEARCH (passive) 5.563A 5.563B		
			238–1,000 GHz (EHF)		
238–240 FIXED FIXED-SATELLITE (space-to-Earth) MOBILE RADIOLOCATION RADIONAVIGATION RADIONAVIGATION-SATELLITE			238–240 FIXED FIXED-SATELLITE (space-to-Earth) MOBILE RADIOLOCATION RADIONAVIGATION RADIONAVIGATION-SATELLITE		

(Continued)

TABLE F.1 (*Continued*)
Worldwide Frequency Allocations
Table of Frequency Allocations

238–1,000 GHz (EHF)

International Table			United States Table		FCC Rule Part(s)
Region 1 Table	Region 2 Table	Region 3 Table	Federal Table	Non-Federal Table	
240–241 FIXED MOBILE RADIOLOCATION			240–241 FIXED MOBILE RADIOLOCATION		
241–248 RADIO ASTRONOMY RADIOLOCATION Amateur Amateur-satellite 5.138 5.149			241–248 RADIO ASTRONOMY RADIOLOCATION 5.138 US342	241–248 RADIO ASTRONOMY RADIOLOCATION Amateur Amateur-satellite 5.138 US342	ISM Equipment (18) Amateur Radio (97)
248–250 AMATEUR AMATEUR-SATELLITE Radio astronomy 5.149			248–250 Radio astronomy US342	248–250 AMATEUR AMATEUR- SATELLITE Radio astronomy US342	Amateur Radio (97)
250–252 EARTH EXPLORATION-SATELLITE (passive) RADIO ASTRONOMY SPACE RESEARCH (passive) 5.340 5.563A			250–252 EARTH EXPLORATION-SATELLITE (passive) RADIO ASTRONOMY US74 SPACE RESEARCH (passive) 5.563A US246		

(*Continued*)

TABLE F.1 (Continued)

Worldwide Frequency Allocations

Table of Frequency Allocations			238–1,000 GHz (EHF)		
International Table			United States Table		FCC Rule Part(s)
Region 1 Table	Region 2 Table	Region 3 Table	Federal Table	Non-Federal Table	
252–265 FIXED MOBILE MOBILE-SATELLITE (Earth-to-space) RADIO ASTRONOMY RADIONAVIGATION RADIONAVIGATION-SATELLITE 5.149 5.554			252–265 FIXED MOBILE MOBILE-SATELLITE (Earth-to-space) RADIO ASTRONOMY RADIONAVIGATION RADIONAVIGATION-SATELLITE 5.554 US211 US342		
265–275 FIXED FIXED-SATELLITE (Earth-to-space) MOBILE RADIO ASTRONOMY 5.149 5.563A			265–275 FIXED FIXED-SATELLITE (Earth-to-space) MOBILE RADIO ASTRONOMY 5.563A US342		
275–1000 (Not allocated) 5.565			275–1000 (Not allocated) 5.565		Amateur Radio (97)

Source: Reproduced from International Telecommunication Union (ITU) sources.

Index

For Product Safety Concerns and Information please contact our
EU representative GPSR@taylorandfrancis.com Taylor & Francis
Verlag GmbH, Kaufingerstraße 24, 80331 München, Germany